ST  AND

ST. W9-APL-037)686

Advances in Soil Science

SOIL BIOLOGY: EFFECTS ON SOIL QUALITY

Edited by

J.L. Hatfield
B.A. Stewart

LEWIS PUBLISHERS
Boca Raton   Ann Arbor   London   Tokyo

Library of Congress Cataloging-in-Publication Data

Soil biology : effects on soil quality / edited by J.L. Hatfield, B.A. Stewart.
      p. cm. — (Advances in soil science)
    Includes bibliographical references and index.
    ISBN 0-87371-927-1
    1. Soil biology. 2. Soils—Quality.   I. Hatfield, Jerry L.
II. Stewart, B. A. (Bobby Alton), 1932–   .   III. Series: Advances in soil science
(Boca Raton, Fla.)
QH84.8.S6315   1993
574.909'48—dc20   93-27391
                                                                  CIP

This book contains information obtained from authentic and highly regarded sources. Reprinted material is quoted with permission, and sources are indicated. A wide variety of references are listed. Reasonable efforts have been made to publish reliable data and information, but the author and the publisher cannot assume responsibility for the validity of all materials or for the consequences of their use.

Neither this book nor any part may be reproduced or transmitted in any form or by any means, electronic or mechanical, including photocopying, microfilming, and recording, or by any information storage and retrieval system, without prior permission in writing from the publisher.

CRC Press, Inc.'s consent does not extend to copying for general distribution, for promotion, for creating new works, or for resale. Specific permission must be obtained in writing from CRC Press for such copying.

Direct all inquiries to CRC Press, Inc., 2000 Corporate Blvd., N.W., Boca Raton, Florida 33431.

© 1994 by CRC Press, Inc.
Lewis Publishers is an imprint of CRC Press

No claim to original U.S. Government works
International Standard Book Number 0-87371-927-1
Library of Congress Card Number 93-27391
Printed in the United States of America 1 2 3 4 5 6 7 8 9 0
Printed on acid-free paper

Preface

Soil quality has become a focal point for attempts to quantify changes in the soil caused by various soil management practices. At the present time, however, we lack measurements which can provide that quantification and detailed understanding of the processes which are occurring within the soil system. We do know that the soil is a complex system full of life and teeming with biological resources. If we are to achieve our goal of improving the soil, we need to understand the soil biological system and its relationship to soil quality.

A high quality soil is thought to include the elements of improved soil aggregation, enhanced water holding capacity, rapid infiltration, increased nutrient availability, extensive rooting depth, increased soil organic matter, reduced pesticide leaching, and resistance to compaction. The challenge is to describe how these factors relate to one another or how each of these can be measured to show how the soil changes under different management practices. Scientists are actively searching for the combination of factors which can be used to describe soil quality. Soil biology is a cornerstone in developing these indices and yet the understanding of the underground biological system is in the early stages of maturity as a scientific discipline. There are many areas of soil biology which need to be studied from a holistic viewpoint and in concert with other disciplines. It is our intent that these chapters would spark some of that interaction.

The biological system present in the soil ranges from the minute to the large mammals, e.g., gophers and moles. One can easily see the effect of large burrows left by some of the mammals. The effect of the microscopic bacteria and fungi are less dramatic but may be the key to improving soil quality. The intent of this volume was to bring together experts who have been working in the area of soil biology to summarize the research progress and propose areas in which additional research could enhance our understanding of the soil biological system.

This volume is a result of the 1991 workshop on Long-Term Soil Management sponsored by the National Soil Tilth Laboratory. Our goal in this workshop was to discuss the current issues in soil biology in context to soil quality and develop a linkage among disciplines in the pursuit of a more complete understanding of the soil system. This Workshop was attended by more than 50 scientists from the United States who unselfishly shared their thoughts and concepts. The chapters which are contained in this volume represent the overview presentations which were used to spur the discussions. We are fortunate to have these individuals summarize their thoughts and ideas in these chapters. Their challenges to the scientific community are quite evident.

The chapters range from the ecology of microbes in different farming systems to the dynamics of the rhizosphere and mycorrhizae, earthworms and other soil fauna, nitrogen cycling, and microbial degradation of pesticides. The range of topics is large and presents an overview of the subject. There is a great deal known about the soil biological system; however, we still have much to learn if we are to improve our soil resources.

Each author presents a challenge to the reader and we hope that these chapters will present a challenge to you. Soil biology is a fascinating area and as we increase our knowledge and understanding, the impact of this understanding will be quite evident in improved soil management practices and environmental quality.

J.L. Hatfield
B.A. Stewart

Contributors

Edwin C. Berry, National Soil Tilth Laboratory, Agricultural Research Service, U.S. Department of Agriculture, Ames, IA 50011, USA

J.W. Doran, U.S. Department of Agriculture, Agricultural Research Service, Soil and Water Conservation Research Unit, Lincoln, NE 68583-0934, USA

D.M. Linn, U.S. Department of Agriculture, Soil Conservation Service, Lincoln, NE 68583, USA

P.D. Milner, U.S. Department of Agriculture, Agricultural Research Service, Soil-Microbial Systems Laboratory, Beltsville, MD 20705, USA

Thomas B. Moorman, U.S. Department of Agriculture, Agricultural Research Service, National Soil Tilth Laboratory, Ames, IA 50011-4420, USA

J.L. Smith, U.S. Department of Agriculture, Agricultural Research Service, Land Management and Water Conservation Research Unit, Pullman, WA 99164-6421, USA

S.F. Wright, U.S. Department of Agriculture, Agricultural Research Service, Soil-Microbial Systems Laboratory, Beltsville, MD 20705, USA

Contents

Microbial Ecology of
Conservation Management Systems

J.W. Doran and D.M. Linn

I. Introduction

Fertile and productive soils are a primary resource of civilization and an essential link in the cycle of life. Terrestrial plants use the energy from sunlight and basic nutrients in the soil and atmospheric environments to synthesize food compounds used by higher life forms. Microorganisms, the unseen citizens of the soil, control soil productivity by recycling the carbon, nitrogen, and other mineral containing compounds in plant and animal residues to forms once again available for use by plants. The soil microbial community also regulates the production and destruction of environmental pollutants such as nitrous oxide, methane, nitrate, and other biologically toxic compounds. Numerical abundance, fast reproductive rates, diversity of type and metabolic activity, and tolerance to a wide range of environmental conditions are unique characteristics of microbial populations within the soil. Consequently, the soil microbial community possesses an enormous catalytic power and metabolic potential. This is aptly demonstrated by the fact that the majority of carbon immobilized each

This chapter was prepared by U.S. government employees as part
of their official duties and legally cannot be copyrighted.

year through photosynthesis by terrestrial plants is returned to the atmosphere as CO_2 by the respiratory activity of heterotrophic soil microorganisms (Bolin, 1970; Coleman and Sasson, 1978).

Agricultural management practices greatly influence the soil ecosystem and the stability and diversity of the microbial community. Tillage and crop residue management practices are major determinants of soil temperature and water regimes and the spatial and temporal availability of energy and nutrients to microorganisms. Changes in tillage and residue management are usually accompanied by associated changes in the availability of essential nutrients to crop plants, soil structure and erodibility, incidence of plant disease, and the microbial production and stability of compounds regulating plant growth. Thus, our understanding of the ecosystem characteristics associated with soil management practices and their influence on microbial activities is critical to the successful maintenance of soil productivity.

A. Tillage Traditions

Historically, soil tillage has played a central role in agricultural productivity. Tillage loosens the soil and prepares a seedbed for germinating seeds and controls the growth of weeds that compete with crop plants for needed water and nutrients. In the early 1700s Jethro Tull, an English farmer, recognized the additional value of tillage for increasing soil nutrient supplies to crop plants. Tull's practical innovations and forceful writing contributed to the commonly held belief of English farmers that frequent stirring of the soil benefited the crop (Pereira, 1975).

Agricultural development in America during the early colonial period resulted in clearing of large areas of forest in the Eastern United States. In the American climate of hotter summers and extremes in rainfall, the traditions of tillage brought from Europe resulted in tremendous losses of topsoil through erosion. Early conservationists warned of the dangers to soil productivity posed by indiscriminate tillage and plowing (see McDonald, 1941). However, uncleared land was abundant and as soil productivity declined many early settlers abandoned their farms and moved to virgin land in the western wilderness.

The moldboard plow was an essential tool of the early pioneers in settling the prairies in the Central and Western United States. It permitted them to create a soil environment in which grain crops could thrive sufficiently to meet the needs of these farmers and an increasing population. The unseen cost of their labors was the oxidation of soil organic matter and depletion of soil fertility reserves. Within the life spans of many of these early farmers, the organic matter level of prairie surface soils declined 40 to 60%. The decrease of soil organic matter, coupled with drought conditions, resulted in an alarming loss of soil by water and wind erosion during the 1930s.

B. Conservation Management

Concern over soil erosion and increasing pressure to farm land too steep or dry for conventional practices led to the development of reduced tillage and residue management systems that conserved crop residues on the soil's surface. Prior to the 1960s these conservation management systems generally involved some form of reduced or subsurface tillage (McCalla and Army, 1961; Triplett, 1982). The introduction of selective herbicides in the 1950s, however, ushered forth the era of weed control without tillage.

Within the past two decades the merits of reduced and no-tillage management systems were recognized both in the United States (Allmaras and Dowdy, 1985; Phillips and Phillips, 1984; Sprague and Triplett, 1986), the United Kingdom (Davies and Cannell, 1975), Europe (Bakermans and DeWit, 1970) and the tropics (Greenland, 1975). Increased interest in conservation tillage arises from the advantages these systems offer over conventional tillage practices. Advantages include: reduced soil erosion losses and increased use of land too steep for farming by conventional tillage; improved timing for planting and harvesting and increased potential for double cropping; conservation of soil water through decreased evaporation and increased infiltration; and a reduction in fuel, labor, and machinery requirements.

Adoption of reduced tillage techniques in the United States has increased steadily during the last three decades. In a preliminary technological assessment, the USDA (1975) predicted that by the year 2000 over 80% of U.S. cropland would be under some type of reduced tillage management of which 45% would be no tillage. In 1980, Phillips et al. estimated that 65% of the corn and soybeans in the southern corn belt would be grown using the no-tillage system by the year 2000. Between 1973 and 1981, the U.S. crop area in minimum tillage increased by 20 million ha or 125% (Christensen and Magleby, 1983). Expansion of no-tillage production during the same period was 1.5 million ha or 78%. Conventional-tillage acreages increased by only 1% during this period, but in 1981 it was still the most used practice on the majority (65%) of U.S. cropland. Estimates and projections for conservation tillage usage in the U.S. were complicated by vague definition of terms. This difficulty was partially overcome by the adoption in 1983 of a definition of conservation tillage as "Any tillage and planting system in which at least 30% of the soil surface is covered by plant residue after planting or where at least 1,120 kg/ha of flat small grain residue are on the soil surface during critical periods for wind erosion" (CTIC, 1983). Using this definition, Schertz (1988) reported that conservation tillage acreages had increased from 2% of the total planted cropland in 1968 to 33% in 1986 and projected that conservation tillage would be practiced on 63 to 82% of total U.S. planted cropland by the year 2010.

Adoption of conservation tillage management practices is, however, hampered by certain disadvantages when compared with conventional systems. These include: cooler soil temperatures, which in temperate and cold climates impede germination and early crop growth; increased potential for insect and disease damage to crops resulting from residue accumulation at the soil surface; and need for more precise management of soil fertility and weed control in achieving desired yields.

C. Purpose of Paper

The purpose of this paper is to provide a current assessment of the soil environment under conservation management practices as related to microbial populations, activities, and processes related to plant growth. The beneficial effects of conservation management on increased soil water storage, reduced erosion, and conservation of organic matter and productivity are well recognized. Crop responses to conservation management, however, vary considerably over ranges of climate, soils, time, and specific management situations.

We have used an empirical approach to defining the soil microenvironment with conservation management to identify the limiting factors and tolerance ranges for microbial activity and establish a basis for identifying the mechanisms responsible for the observed biological responses. Discussion centers primarily on surface soil characteristics (0 to 30 cm) since this is the major zone of microbial activity. To evaluate the influence of reduced tillage and residue management practices on the soil ecosystem, we have relied heavily on comparison with conventional tillage. Therefore, most comparisons are presented for the extremes of tillage: no tillage, where soil is disturbed only during planting, and conventional tillage, where soil is inverted with the moldboard plow followed by one or more secondary tillage operations to prepare a seedbed and incorporate crop protection chemicals and fertilizers. For overviews of general aspects of conservation tillage the reader is referred to the following proceedings, special publications, and review articles (Baeumer and Bakermans, 1973; Logan et. al., 1987; McCalla and Army, 1961; Phillips and Phillips, 1984; Plant Protection Limited, 1973,1975; Power, 1987; Sprague and Triplett, 1986; Soil Conservation Soc. Am. 1973, 1983; and Unger and McCalla, 1980).

II. Management and the Soil Micro-environment

The soil ecosystem represents an intricate balance between living and non-living components. Biological responses to changes in the physical and chemical environment predominantly relate to biological needs for energy, an energy sink (usually oxygen), water, nutrients, suitable temperature and living space, and the absence of harmful conditions. The magnitude and frequency of fluctuations in these requirements largely govern the activities and predominance of microorganisms and plants.

Agricultural production systems involving tillage, residue, and crop management practices directly, and indirectly, affect soil environmental factors that control the growth and activity of plants and microorganisms. These two groups

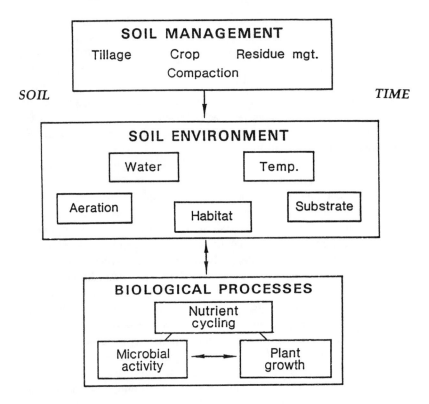

Figure 1. Relationship between soil management and the soil microenvironment as it determines biological processes influencing nutrient cycling and interactions between microbial activity and plant growth.

of organisms, in turn, directly influence nutrient cycling in soils (Figure 1). Management practices which influence the placement and incorporation of plant and animal residues also control the accessibility of these substrates to soil microflora and fauna. Furthermore, residue placement indirectly influences such soil environmental factors as water and aeration status, temperature, and the relative predominance of certain organisms. These environmental factors, within the constraints set by climate, soil type, vegetation, and time, play a major role in controlling the biological processes that determine microbial activity and the cycling of N and other elements. The degree of soil disturbance associated with tillage influences the aforementioned environmental factors and the spacial habitat for biological activity through effects on soil bulk density, pore continuity and size distribution, and aggregation. In trying to sort out the complex interactions between management and nutrient cycling, the synchrony of plants and microorganisms responding to changes in the soil environment is important. Moreover, the dynamics of these interactions become a major consideration where organic matter reserves are augmented *in situ* through use of green manure and cover crops. Evaluating the effects of agricultural management practices on microbial activity, nutrient cycling, and plant growth is further complicated by interactions between residue placement and soil disturbance. The major influence of each, however, appears predominantly related to changes in the distribution of substrates, organisms, and alteration of the soil biophysical environment.

Tillage affects a multitude of properties which define the soil microenvironment. Tillage directly affects soil porosity and placement of crop residues, which in turn affect soil physical, chemical, and biological properties. Porosity influences the air- and water-holding capacity of soil. Crop residue placement affects surface soil temperature, evaporation rate and soil water content, nutrient distribution, and the positional availability of organic C and N compounds that serve as microbial substrates. Collectively, these factors define the soil microenvironment, which in turn determines the relative predominance of specific microorganisms and their activities. Through this chain of events, tillage and residue management practices influence the physical, chemical, and biological components ultimately regulating soil productivity and environmental quality. Crop productivity is therefore both directly and indirectly influenced by the soil micro-environment determined by type and degree of tillage.

The degree to which soil environmental components are changed by management practices and their relative importance to biological productivity is largely determined by the state factors of soil type, temporal stage, characteristic crop or vegetation, and climate, as described by Jenny (1980).

III. Tillage Management and Microbial Activity

A. Nutrient Cycling

Of the many biological processes influenced by tillage and residue management, mineralization/immobilization are most important to nutrient cycling. Of particular importance is the potential for decreased mineralization of soil organic N and increased immobilization of surface fertilizer N applications with reduced or no-tillage systems in which residues and microorganisms are concentrated near the soil surface (Doran, 1980a; Follett and Schimel, 1989; Kitur et al., 1984). When crop residues are routinely burned before planting there may be little or no difference in N immobilization between no-tilled and plowed soils (Dowdell and Crees, 1980). However, where crop residues are retained on the soil surface and crops are planted directly into a chemically-killed cover crop with a high C/N ratio, immobilization of soluble N can be much greater in no-tillage than plowed surface soils (Rice and Smith, 1984).

Table 1. Comparison of soil properties and microbial parameters for no-tillage relative to plowing across six U.S.A. locations

| Soil Depth (cm) | Ratio, No-tillage versus Plow | | | | |
	Soil Water	Total Organic C	Total Kjeldahl N	Mineralizable N	Microbial Biomass
0-7.5	1.3	1.4	1.3	1.4	1.5
7.5-15	1.1	1.0	1.0	1.0	1.0
15-30	1.1	0.9	1.0	0.9	1.0

(after Doran, 1987)

The stratification of crop residues, organic matter, and soil organisms with no-tillage management slows the recycling of N as compared with conventional tillage operations (Fox and Bandel, 1986; House et al., 1984). Surface soil levels of organic matter, microbial biomass, and potentially mineralizable N are all significantly higher with no tillage as compared with moldboard plow tillage (Table 1). As a result of increased residue and organic matter levels and more optimal water status, greater microbial biomass in surface no-tillage soils is associated with greater reserves of potentially mineralizable N. However, the absolute magnitude of the differences in biomass and potentially mineralizable N may be greater in warm humid climates where plant production and use of cover crops is also greater (Ayanaba et al., 1976; Doran, 1987). Increased microbial biomass is also often associated with higher density and activity of plant roots in surface no-tillage soils (Carter and Rennie, 1984; Lynch and Panting, 1980).

Interactions between microbial activity and mineralization of soil organic N are often controlled by environmental factors (Duxbury et al., 1989). Predicting how environmental conditions under reduced tillage affect net mineralization is complicated by the contrasting effects of increased water and reduced temperature which vary during the growing season and across climates (Doran and Smith, 1987; Fox and Bandel, 1986). In the early growing season cooler and wetter soil conditions associated with reduced tillage may result in lower microbial activity and mineralization compared with conventional tillage. Mineralization later in the growing season, when temperatures are more favorable for biological activity, may be greater under reduced tillage as a result of more optimal soil water contents. Also, greater microbial biomass in no-tillage surface soils during the growing season may serve as a sink for immobilization of N. Higher soil microbial biomass levels under no-tillage production of wheat and corn have been related to greater immobilization of fertilizer N as compared with plowing or shallow tillage (Carter and Rennie,

1984; Rice et al., 1986). Development of fertilizer application methods such as banding, split application, and spoke-wheel injection have greatly improved N-use efficiency in no-till management systems (Mengel et al., 1982; Timmons et al., 1987; Touchton and Hargrove, 1982). These technological adaptations provide a good example of the importance of understanding soil ecology in development of sustainable and efficient agricultural production systems (Hardin, 1989; Mowitz, 1991; Wanzel, 1988).

Besides vertical stratification with depth in soil, horizontal distributions of crop residues across the soil surface can greatly influence soil biological activities and chemical properties. In drier climates the availability and uptake of N from soil organic matter, crop residues, and applied fertilizer can be increased by the presence of surface crop residues creating a soil environment more favorable for microbial activity, N mineralization, and plant growth (Power et al., 1986). In the subhumid Central United States, surface soil microbial populations between crop rows with zonal tillage management practices, such as ridge tillage, can be 2 to 40 times greater than within crop rows (Doran, 1980a). Between crop rows, where crop residues are concentrated with this management system, soil water contents, substrate supply, and soil pH are more optimal for microbial activity than within crop rows. We found up to 145 kg NO_3-N/ha accumulated in surface soil between rows due, in part, to increased mineralization/nitrification resulting from more optimal soil water and pH regimes. This pool of plant-available N, however, was located in dry soil and unavailable for uptake by crop roots during much of the summer growing season; consequently it was susceptible to loss after harvest through leaching or denitrification when soil conditions were generally wetter. This illustrates the importance of spacial as well as temporal synchronization of nutrient availability with crop growth and root activity.

Degree of soil tillage can profoundly influence oxygen dependent microbial transformations of C and N (Linn and Doran, 1984b). The relative predominance of anaerobic microorganisms in the surface of no-tillage soils is often greater than that of aerobic microorganisms as compared with plowed soils (Table 2). This difference is often associated with wetter more compact conditions under reduced tillage giving rise to less aerobic conditions with no tillage resulting from greater water content and/or decreased soil porosity (Linn and Doran, 1984b). Greater soil water content and reduced air-filled porosity, particularly after rainfall or irrigation, may enhance denitrification and gaseous losses of N in no-tillage as compared with plowed soils (Aulakh et al., 1991a; Aulakh et al., 1982; Rice and Smith, 1982). However, as shown in Figure 2, the potential for less aerobic conditions with reduced tillage is site specific and depends on climate, soil porosity, drainage characteristics, and the quantity of crop residues maintained on the soil surface (Doran et al., 1987b). In drier climates greater potential for denitrification with reduced tillage, as indicated by higher supplies of organic substrates and NO_3-N and greater numbers of denitrifying organisms, may rarely result in significant losses of N because soil aeration is only occasionally limited by excessive water contents (Doran and Smith, 1987).

Table 2. Ratio of microbial populations between no-till and moldboard plow tillage, average of seven U.S.A. locations

Microbial Group	Ratio, No Till/Plow by depth (cm)		
	0-7.5	7.5-15	0-15
Aerobes			
Fungi	1.4	0.6	1.0
Bacteria	1.4	0.7	1.1
Nitrifiers	1.0	0.5	0.8
Anaerobes			
Denitrifiers	2.7	1.9	2.3
Facultative	1.3	1.0	1.1
Obligate	1.3	1.1	1.2

(adapted from Doran 1980b; Linn and Doran, 1984a)

Significant loss of N from denitrification in reduced tillage soils is apparently limited to poorly or imperfectly drained soils (Blevins et al.,1984).

B. Plant Disease

Although maintenance of crop residue on the soil surface from conservation tillage practices can offer a variety of advantages over conventional tillage, disadvantages can also be encountered (Phillips et al., 1980). For example, the presence of surface crop residues can influence the outbreak and severity of many crop diseases. Extensive reviews regarding crop diseases in relation to conservation tillage and surface crop residues are presented by Boosalis et al. (1981, 1986); Cook et al. (1978); and Sumner et al. (1981). The reader is referred to these reviews for discussions concerning effects of conservation tillage practices on specific diseases and disease organisms for a wide variety of crops.

The influence of conservation tillage on plant disease encompasses the complete spectrum of possibilities, ranging from decreased to increased incidence depending upon the crop, cultural, and climatic factors (Boosalis et al., 1981, 1986). Although a number of factors may be involved, the primary factor is the presence of crop residues on the soil surface. Nearly all plant pathogenic fungi and bacteria require plant residues to successfully complete some aspect of their life cycle (Cook et al., 1978). As a result, the most frequently encountered diseases associated with conservation tillage systems have been fungal or bacterial in origin (Boosalis et al., 1981). Residues maintained

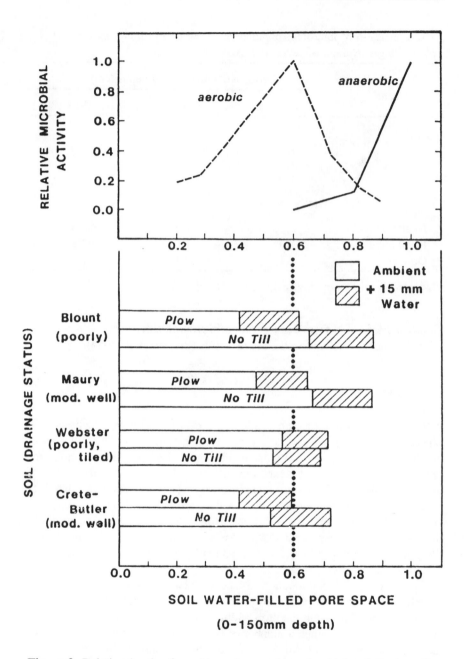

Figure 2. Relative levels of aerobic and anaerobic microbial activity in soil as determined by degree of water-filled pore space as influenced by addition of water, tillage management, and soil drainage status. (After Mielke et al., 1986.)

on the soil surface can thus directly influence the incidence of crop disease by not only providing a suitable habitat for survival and reproduction of plant pathogens but also by serving as an inoculum source for succeeding crops (Boosalis and Doupnik, 1976).

Surface residues may also affect the incidence of crop disease indirectly. Changes in the soil/plant environment, by comparison to conventional tillage practices, are significant in this regard. Among the changes brought about by surface residues that may create conditions favorable for development of disease are those involving physical and chemical soil properties. Changes in the soil/plant environment governed by these properties can alter either host or pathogen nutrition, growth and configuration of plant roots, or host susceptibility. Changes among any or all of these parameters have been cited as particularly important to the ability of plant pathogens to infect a host crop (Cook et al., 1978).

Soil temperature and water content are among the most important physical factors affected by surface residues. Residue covered soils are often cooler, especially early in the growing season, and wetter than conventionally tilled soils (Baeumer and Bakermans, 1973; Phillips et al., 1980). This combination of lower temperatures and greater availability of soil water is conducive to the development of damping-off, seedling blight, and root disease (Boosalis et al., 1981, 1986). In cases where plant pathogens are obligate aerobes the more dense and often wetter soil conditions of reduced-tillage management can result in decreased disease incidence. Although measurements of soil water status were not made, Rothrock (1987) suggested that lower incidence of take-all disease of wheat with no tillage in the Southeastern United States may result from a cooler, wetter environment as compared with conventional tillage. Heritage et al. (1989) demonstrated that the infection of wheat and growth of the take-all fungus, *Gaeumannomyces graminis,* was severely restricted at soil water-filled pore space values exceeding 80%. As discussed earlier, the probability of achieving this degree of soil saturation is much higher with reduced-tillage than conventional tillage management systems.

The chemical environment of the soil can also change with conservation tillage practices. Surface residues are known to influence the rate at which nutrients become available to crop plants (Power and Legg, 1978), the distribution of nutrients within the soil profile (Baeumer and Bakermans, 1973), and the tendency for the development of an acidic soil pH in the absence of an adequate liming schedule (Blevins, et al., 1978). Changes in these soil parameters are thought important to the survival and growth of soil-borne pathogens under conservation tillage conditions (Cook et al., 1978; Sumner et al., 1981). Alternately, Baker and Cook (1974) proposed that biological control of soil-borne pathogens, through antagonistic and competitive interactions with non-pathogens, may also occur with conservation tillage because of the stimulatory effect these practices have on surface soil microbial populations.

Fortunately control and prevention of plant diseases associated with conservation tillage practices can be achieved through a variety of means.

Foremost among these is a requirement on the part of the grower for conscientious and careful management. The use of disease resistant crop varieties and crop rotations, instead of monocropping practices, are two effective means of disease prevention (Boosalis et al., 1981, 1986; Cook et al., 1978, Rothrock and Cunfer, 1986). Controlling weed growth is also important since many weeds also serve as host for viruses or insect vectors of pathogens, or provide a habitat for the survival, growth, or multiplication of pathogens (Cook et al., 1978). The success of weed control and crop rotations in disease prevention with conservation tillage has best been demonstrated with the ecofallow system (Smika and Wicks, 1968). This practice, used primarily in the Great Plains region of the United States, employs weed control through sweep tillage and/or herbicide application with a wheat-sorghum or corn-fallow rotation. Although primarily designed as a crop-management practice to conserve soil moisture, the ecofallow system can also substantially increase the grain yield of sorghum by decreasing stalk rot losses (Doupnik et al., 1975). The crop rotation aspect of the ecofallow system is a major factor in decreasing stalk rot losses since each succeeding crop is planted into the residue of a different crop.

Other factors which can reduce disease incidence in conservation tillage systems include: development of crop varieties specifically adapted to the conservation tillage ecosystem (Boosalis et al. 1981, 1986); timing, type, and amount of fertilizer application (Huber and Watson, 1974); biological control (Baker and Cook, 1974); and chemical control. As pointed out by Boosalis et al. (1981) the concept of integrated pest management (Apple, 1973) holds particular promise for the control of crop disease in conservation tillage systems because of its incorporation of biological, chemical, and cultural techniques to control crop pathogens and, as a result, the outbreak and severity of crop diseases.

C. Phytotoxins

Under certain conditions, the microbial decomposition of crop residues in or on the soil can result in the release of compounds that are toxic to crop plants (Cochran et al., 1977; Guenzi and McCalla, 1962; Lynch, 1977; McCalla and Duley, 1948; Patrick and Koch, 1958). Crops exposed to phytotoxins exhibit a range of symptoms, including: reduced germination, seedling growth inhibition or death, chlorosis, abnormal development of roots or foliage, and reduced tillering (Cochran, et al., 1977; Kimber, 1967; McCalla and Haskins, 1964; McCalla and Norstadt, 1974; Rice, 1984). In addition to the direct effects on plant development, phytotoxins can also predispose plant roots to attack by soil-borne pathogens (Linderman and Toussoun, 1968; Lynch, 1987; Patrick and Koch, 1963; Toussoun and Patrick, 1963).

The release of phytotoxic substances from crop residues occurs by at least two mechanisms. First, phytotoxins can be released directly from fresh or unweathered residues during early stages of microbial decomposition (Cochran et al.,

1977; Guenzi and McCalla, 1962; Guenzi et al., 1967; Patrick, 1971; Toussoun, et al., 1968). Second, decomposing plant residues can stimulate the growth and activity of certain microorganisms which, in themselves, are capable of producing substances toxic to plant growth (Elliott, et al., 1980; Lynch, 1977, 1978; McCalla and Haskins, 1964; Norstadt and McCalla, 1963, 1969; Tang and Waiss, 1978).

A number of phytotoxins, both those released from decaying residues or produced by microorganisms, have been isolated and identified (McCalla and Haskins, 1964). Included among those phytotoxins released from residues during decomposition are a variety of water-soluble organic acids, including phenolic, aromatic, and short-chain fatty acids (Guenzi and McCalla, 1966; Patrick, 1971; Toussoun et al., 1968). Phytotoxins produced by microorganisms include those of fungal origin, such as patulin (Norstadt and McCalla, 1963) and oxalic acid (McCalla et al., 1963) or metabolic by-products of microbial fermentation such as acetic acid (Harper and Lynch, 1981; Lynch, 1977, 1978). A number of volatile compounds released or produced from decomposing residues can also affect select groups of microorganisms (Bremner and Bundy, 1974; Menzies and Gilbert, 1967; Smith, 1973, 1976).

Phytotoxin development in soil is favored by cool, wet or anaerobic conditions (Cochran et al., 1977; Lynch, 1978, 1987; McCalla, et al., 1963; Patrick, 1971). Such conditions are often prevalent early in the growing season in temperate regions, and particularly so with soils subject to conservation tillage practices. As a result, the occurrences of yield reductions with crops grown under conservation tillage systems have, in some instances, been attributed to phytotoxins (McCalla and Army, 1961; McCalla and Haskins, 1964). Fortunately, phytotoxin production, as well as toxicity, is generally a short-lived phenomenon. Primary among the reasons for the short-lived nature of many phytotoxic compounds is their vulnerability to rapid microbial degradation (Patrick, 1971). In addition to microbial degradation, phytotoxin production is limited to fresh or unweathered residues and conditions favorable for microbial proliferation (Cochran et al., 1977; Guenzi et al., 1967; Harper and Lynch, 1981; Kimber, 1967, 1973).

The benefits derived from conservation tillage practices, particularly soil and water conservation, should not be overshadowed by the potential for phytotoxin development resulting from the use of such practices. The reason benefits outweigh disadvantages is that residue and crop management techniques are available, or being developed, which are effective in minimizing the production or duration of phytotoxins in the field. Particularly important is an awareness that the greatest potential for phytotoxin production occurs with fresh residues and cool, wet soils. Hence, planting times can be altered to better avoid these conditions or, as suggested by Lynch and Elliott (1983), intentionally creating conditions in the field to bring about an accelerated decomposition of residues prior to planting. The use of proper tillage implements and planters can also reduce phytotoxicity problems. Planters designed to cut through and clear residues away from the seed row can be particularly effective in controlling phytotoxicity (McCalla and Norstadt, 1974; Woodruff et al., 1966).

The unique ecology of conservation tillage systems and the associated production of phytotoxins from decomposing cover crops has aided the development of sustainable management systems that use cover crops for erosion control, nutrient retention, and weed control. Putnam and DeFrank (1983) found selective control of several annual weed species was enhanced by decomposition of grass or cereal grain cover crops. The allelopathic activity of rye (*Secale cereale* L.) as a winter cover crop in controlling annual weeds is also enhanced by no-tillage management (Barnes and Putnam, 1983, 1986).

IV. Sustainable Management Systems

Historically, agricultural systems evolve as we manage nature to meet our food and fiber needs and adjust for the environmental consequences of altering natural ecological balances. Defining sustainable management systems for agriculture, however, is complicated by the need to consider their utility to humans, efficiency of resource use, and favorable balance with the environment (Harwood, 1990). Major technological revolutions in the United States during this century have dramatically influenced production and sustainability in agriculture. Agricultural mechanization at the turn of the century enabled massive cultivation of fertile virgin prairie soils in America's breadbasket. However, without return of N and other nutrients, the organic matter level in Great Plains soils declined dramatically and agricultural sustainability was threatened by drought and unprecedented soil erosion in the 1930's. Conservation tillage and green manuring practices were introduced to reduce erosion, microbial oxidation of organic matter, and rebuild soil fertility. After World War II, inexpensive and abundant fertilizer-N decreased the role of N-fixing legume cover crops in cropping systems (Power and Papendick, 1985). Increased cultivation and monoculture production of cash grain crops, development of higher yielding varieties, and greater reliance on chemical and energy inputs increased crop yields 2-3 fold in the 40 years following World War II. These yield increases, however, were also associated with declines in native soil fertility, reduced profitability from greater capital inputs, and increased soil erosion and environmental loading of certain agrichemicals in parts of the United States. At present, agriculture is considered by some, the largest areal contributor to nonpoint source water pollution in the United States (National Research Council, 1989). Thus we are once again faced with adopting agricultural management approaches which will balance adequate production levels with acceptable environmental quality.

Table 3. Soil productivity distributions between alternative management using legumes in rotation and conventional management using fertilizers and herbicides 2 years after conversion from conventional management

Biomass Component	Dry Matter Yields		Nitrogen Yields	
	Organic Legume N	Conventional Chemicals	Organic Legume N	Conventional Chemicals
	----kg/ha----		----kg N/ha----	
Maize (grain, stover, roots)	12,910	22,500	127	242
Weeds (tops + roots)	6,280	210	52	2
Microbial biomass (top 30 cm soil)	2,620	1,980	157	119
Totals	21,810	24,740	335	363

(after Doran et al., 1987a)

A. Chemical Inputs/Cropping Systems

Alternative crop management systems that will reduce off-farm inputs are being developed to offset rising production costs, decrease environmental and health hazards associated with the use of certain agricultural chemicals, and maintain soil productivity levels. Such systems commonly use animal wastes and legume and green-manure crops in rotation with cash grain crops to reduce use of inorganic fertilizers and pesticides. These alternative systems, however, may initially result in lower cash grain yields due to an altered biological environment, conservation of C and N in the soil/plant system, increased competition from weeds, and user adjustment to new management techniques (Doran and Werner, 1990).

In the humid Eastern United States, Liebhardt et al. (1989) reported maize yields from organic management systems (using animal manure and legumes as N sources) during the second year of conversion from conventional management were 40 to 50% lower than conventional systems where herbicides and fertilizer were used. Decreased production resulted primarily from insufficient available soil N and competition from weeds. As illustrated by the data in Table 3, increased dry matter and nitrogen yields in maize with conventional management early balanced increases in weeds and microbial biomass in the low-input organic management system. The higher overall N yields of conventional management reflected the additional application of 123 kg/ha of fertilizer nitrogen, assuming a fertilizer N recovery of about 25%. Thus, lower above-ground crop yields with organic management during the transition period were not the result of decreased soil productivity but rather different partitioning into plant and microbial organic N pools which serve as potential nutrient reserves for future crops. As anticipated, there was no measurable change in soil organic matter during the first two years transition from conventional management. Thus, the total ecological soil productivity (above- and below-ground resources) of organic and conventional systems was similar.

In the above research study, maize grain yield in the reduced input cash grain system in the fifth year after conversion averaged 9.53 Mg/ha, which was equal to or greater than maize grain yield in the conventional management system. By this time, the plant/soil ecosystem with organic management had reached a new equilibrium, resulting in less weed pressure and greater turnover of microbial biomass and recycling of available N to grain crops during the growing season. An important change for the organic management treatment was the overseeding of a hairy vetch (*Vicia villosa* Roth.) winter cover crop into the preceding wheat (*Triticum aestivum* L.) crop. Mineralization of the hairy vetch cover crop (above ground matter contained 182 kg N/ha) increased soil nitrate-N levels by 103 kg/ha within one month after incorporation by plowing (Table 4). Soil microbial biomass N and potentially mineralizable N reserves decreased by 38 and 80 kg N/ha, respectively, during this same time period. This illustrates the importance of microorganisms in mediating release of N from cover crop residues and the importance of management in synchronizing soil N availability with time of maximum need by the grain crop. The hairy vetch cover crop was effective in reducing the potential for over-winter leaching losses of nitrate-N. Nitrate levels shortly before planting in the surface 30 cm of soil with hairy vetch averaged only 9 kg N/ha; well below the late spring value of 36 kg N/ha considered minimal for producing an adequate crop of maize in the Northeastern United States (Magdoff et al., 1984).

B. Conservation Management Challenges

Cash grain yields in management systems using legumes for supplemental N, cover crops to reduce erosion and over-winter nitrate leaching, and some tillage to stimulate N mineralization and weed control depend on the farmer's ability to synchronize available soil N supplies with time of maximum crop need. Early season N release from winter cover crops depends mainly on microbial mineralization/immobilization of C and N which is influenced by the C/N ratio of the residue, its degree of incorporation in soil, and soil temperature and water regimes (Aulakh et al., 1991b; Sarrantonio and Scott, 1988; Wagger, 1989).

Table 4. Management and cover crop effects on N cycling in the top 30 cm soil and changes during the maize growing season in the 5th year's conversion from conventional management

Management and Date	Soil Nitrogen Pools		
	Microbial N	Potentially Mineralizable N	Nitrate N
	------------------kg N/ha----------------		
Alternative			
(maize/clover/winter wheat/soybean rotation + vetch)			
April 10	121	1260	9
May 15[a]	113	1260	39
June 12	75	1180	142
July 14	122	1220	99
Conventional			
(corn/soybean rotation with herbicides and fertilizer)			
April 10	92	990	42
May 15	56	950	56
June 12[b]	64	990	83
July 14	103	1020	106

[a] Hairy vetch cover crop (182 kg N/ha) plowed into soil on May 8.
[b] 112 kg N/ha of ammonium nitrate fertilizer sidedressed on June 17
(after Doran et al., 1987a).

Depending on these soil conditions, it may take 1-3 weeks after cover crop incorporation before N release exceeds N immobilization. The farmer managing biological resources to reduce purchased inputs must decide if additional fertilizer N is needed and, if so, how much. Careful management of available soil N will also reduce the potential loss of N through leaching or denitrification during times of the year when the soil is not cropped.

Ridge tillage is a conservation management technique that uses sweeps to till a zone of soil in the row 15 to 20 cm wide and 5 to 10 cm deep at planting time. This creates a desirable seedbed environment and covers crop residue and weeds in the between row area with soil. Subsequent between-row cultivations 2 to 4 weeks after planting promotes mineralization of soil organic residues, controls weeds, and reforms a ridge in the crop row to provide a raised bed for future crops. In temperate climates, vetch and rye cover crops are often used with ridge tillage management without herbicides to provide winter cover and afford weed control through competition and in the case of rye, allelopathy. The cover crops also protect the soil from erosion and reduce the potential for over-winter leaching losses of available N. In the spring, the farmer who doesn't use herbicides or soil tillage until planting is challenged to provide adequate soil N for his grain crop.

The availability of soil N and crop response to ridge tillage with cover crops is greatly influenced by microbial activity and management related changes in the soil physical environment. Cooler, wetter, and presumably less aerobic soil conditions with ridge tillage management during early spring can result in slower crop growth due to less available N. To illustrate the importance of managing the soil environment to optimize conditions for plant growth and microbial activity we will present research results for tillage management comparisons from a diversified farm in Iowa using ridge tillage for erosion control and cover crops to reduce chemical inputs. As shown in Table 5, soil microbial biomass N levels with ridge tillage in the early growing season averaged 34 to 56 kg/ha more than those of conventional tillage without cover crops; mainly due to maintenance of greater levels of crop residue and available microbial substrates near the soil surface. Soil nitrate-N levels in surface soil with ridge tillage averaged 22 kg/ha less than those with conventional disking and for much of the growing season were below the 70 kg/ha level recommended for non-limited growth of maize in this area. Soil nitrate levels for ridge-tillage management were least in the between-row areas where crop residues were greatest and soil density and water content were highest. Also, gaseous loss of soil nitrate-N through microbial denitrification was also greatest in between-row wheel track areas with ridge-tillage management and reached maxima of 0.5 to 1 kg N/ha/d.

Lower early season soil nitrate N levels with ridge-tillage + winter cover crops apparently limited growth and final grain production of maize as compared with disk tillage without a cover crop. Slower growth and development of maize (12-35% less) in ridge tillage + cover crop as compared with disk tillage without cover also resulted from cooler early season soil temperatures, a less aerobic environment, and perhaps competitive water use by the cover crops in a particularly dry year. Environmental soil conditions and residue accumulations with ridge tillage in the between-row areas during early May apparently resulted in the loss of 20 to 28 kg of nitrate-N/ha, presumably through microbial immobilization and denitrification.

Combining use of cover crops with conservation tillage may result in more sustainable agricultural crop production but the unique soil environments created by such practices must be considered in refining management systems for optimal crop growth and nutrient supply. In temperate climates, shortage of available N during the growing season for reduced tillage management with cover crops might be lessened by supplemental fertilizer N at planting or by earlier tillage to aerate and warm soil and to enhance mineralization and release of N from crop residues, microbial biomass, and soil organic matter (Karlen and Doran, 1991).

Table 5. Effect of tillage management and hairy vetch and rye cover crops on the soil environment, N availability, and 1989 maize yields in central Iowa

Soil Property or Crop Yield	Double Disk No Cover Crops	Ridge-Till with Cover Crops	Difference Ridge-Till vs Disk
Pre-Planting			
Soil Temperature (°C) at 5 cm	22	16 to 18	4-6° cooler
Water-filled Porosity (%)	48	50 to 78[a]	less aerobic
Soil Nitrate-N (kg/ha 30 cm)	49	27	22 kg/ha less
Microbial-N (kg/ha 30 cm)	360	335	less microbial N
7 weeks after planting			
Soil Nitrate-N (kg/ha 30 cm)	77	53	N limiting
Microbial-N (kg N/ha)	300	360	more microbial N
Crop N uptake (kg N/ha)	25	14	44% lower
Crop Dry Matter (kg/ha)	700	460	35% lower
Harvest			
Total N Uptake (kg N/ha)	150	124	17% lower
Final Grain Yield (kg/ha)	9850	8910	10% lower

[a] Highest in wetter, more compact between-row wheel track area with ridge-tillage management (after Doran and Smith, 1991)

V. Conclusions

Agriculture is challenged to develop sustainable management systems that provide sufficient crop yields, environmental protection, and conserve limited resources. Increasing the efficiency of conservation management systems requires understanding the complex effects of tillage, residue, and cropping management on the soil environment for microbial activity and plant growth. Understanding the unique ecology of conservation tillage systems in controlling the relative predominance and potential activity of soil microorganisms and plants has aided technological developments to improve the efficiency of agricultural production. These developments, however, have been limited in scope and application by the empirical and "localized" nature of past conservation management research. A systematic, unified research approach is needed to identify the primary factors influencing the performance of conservation management systems with respect to economic crop production and environmental quality.

A major research need is development of a system to catalog the soil physical, chemical, and biological properties most influenced by conservation management and identify how interaction of these properties determines soil biological activity. Definition of primary soil properties and the soil environment for plant growth and microbial activity requires establishment of interpretive guidelines identifying how variations in soil type, climate, and cropping systems influence the importance and interaction of soil properties in controlling and limiting biological activity. The integrated efforts of many scientific disciplines such as Agronomists, Soil Scientists, Microbiologists, Ecologists, Economists, Climatologists, Computer Scientists and Agricultural Engineers is needed to predict performance and modifications of conservation management systems over a range of soil, climatic, and economic conditions. Successful integration of scientific disciplines will require development of unified databases and decision management systems.

References

Allmaras, R.R. and R.H. Dowdy. 1985. Conservation tillage systems and their adoption in the United States. *Soil Tillage Res.* 5:197-222.

Apple, J.L. 1973. Integrated Pest Management: Philosophy and Principles. p. 89-97. In: L.F. Seatz (ed.), *Symposium on Ecology and Agricultural Production.* Univ of Tennessee, Knoxville.

Aulakh, M. S., D.A. Rennie, and E.A. Paul. 1982. Gaseous nitrogen losses from cropped and summer-fallowed soils. *Can. J. Soil Sci.* 62:187-196.

Aulakh, M.S., J.W. Doran, and A.R. Mosier. 1991a. Soil Denitrification - Significance, measurement, and effects of management. *Advances in Soil Science* 18:1-57.

Aulakh, M.S., J.W. Doran, D.T. Walters, A.R. Mosier, and D.D. Francis. 1991b. Crop residue type and placement effects on denitrification and mineralization. *Soil Sci. Soc. Am. J.* 55:1020-1025.

Ayanaba, A., S.B. Tuckwell, and D.S. Jenkinson. 1976. The effects of clearing and cropping on the organic reserves and biomass of tropical forest soils. *Soil Biol. and Biochem.* 8:519-525.

Baeumer, K. and W.A.P. Bakermans. 1973. Zero tillage. *Adv. Agron.* 25:77-123.

Baker, K.F. and R.J. Cook. 1974. *Biological Control of Plant Pathogens.* W.H. Freeman, San Francisco. 433 pp.

Bakermans, W.A.P. and C.T. DeWit. 1970. Crop husbandry on naturally compacted soils. *Neth. J. Agric. Sci.* 18:225-246.

Barnes, J.P. and A.R. Putnam. 1983. Rye residues contribute weed suppression in no-tillage cropping systems. *J. Chem. Ecol.* 9:1045-1057.

Barnes, J.P. and A.R. Putnam. 1986. Allelopathic activity of rye (*Secale cereale* L.). p. 271-286. In: A.P. Putnam and C-S. Tang (eds.), *The science of allelopathy.* John Wiley & Sons, New York.

Blevins, R.L., M.S. Smith, and G.W. Thomas. 1984. Changes in soil properties under no-tillage. p. 190-230 In: R.E. and S.H. Phillips (eds.) *No-tillage Agriculture.* Van Nostrand Reinhold Co., New York.

Blevins, R.L., L.W. Murdock, and G.W. Thomas. 1978. Effect of lime applications on no-tillage and conventionally tilled corn. *Agron. J.* 70:322-326.

Bolin, B. 1970. The carbon cycle. *Scientific Am.* 223:124-132.

Boosalis, M.G. and B. Doupnik, Jr.. 1976. Management of crop diseases in reduced tillage systems. *Bull. Entomol. Soc. Am.* 22:300-302.

Boosalis, M.G., B.L. Doupnik, Jr., and G.N. Odvody. 1981. Conservation tillage in relation to plant diseases. p. 445-474. In: D. Pimentel (ed.), *Handbook of Pest Management in Agriculture, Volume 1.* CRC Press, Inc. Boca Raton, FL.

Boosalis, M.G., B.L. Doupnik, Jr., and J.E. Watkins. 1986. Effect of surface tillage on plant diseases. p. 389-408. In: M.A. Sprague and G.B. Triplett (eds.), *No-tillage and surface-tillage agriculture: the tillage revolution.* John Wiley & Sons, New York.

Bremner, J.M. and L.C. Bundy. 1974. Inhibition of nitrification in soils by volatile sulfur compounds. *Soil Biol. Biochem.* 6:161-165.

Carter, M. R. and D.A. Rennie. 1984. Nitrogen transformations under zero and shallow tillage. *Soil Sci. Soc. Am. J.* 48:1077-1081.

Christensen, L.A. and R.S. Magleby. 1983. Conservation tillage use. *J. Soil Water Conserv.* 38:156-157.

Cochran, V.L., L.F. Elliott, and R.I. Papendick. 1977. The production of phytotoxins from surface crop residues. *Soil Sci. Soc. Am. J.* 41:903-908.

Coleman, D.C. and A. Sasson. 1978. "Decomposer subsystem" pp 609-655 In: A.J. Breymeyer and G.M. Van Dyne (eds.), *Grasslands, systems analyses and man. IBP 19,* Cambridge Univ. Press, Cambridge.

Cook, R.J., M.G. Boosalis, and B. Doupnik, Jr. 1978. Influence of crop residues on plant diseases. p. 147-163 In: W.R. Oschwald (ed.), *Crop Residue Management Systems*. Am. Soc. Agron. Spec. Pub. 31. Madison, WI.

CTIC (Conservation Tillage Information Center). 1983. *National survey of conservation tillage practices*. Fort Wayne, Ind.

Davies, D.B. and R.Q. Cannell. 1975. Review of experiments on reduced cultivation and direct drilling in the United Kingdom, 1957-1974. *Outlook Agri*. 8:216-220.

Doran, J.W. 1980a. Microbial changes associated with residue management with reduced tillage. *Soil Sci. Soc. Am. J*. 44:518-524.

Doran, J.W. 1980b. Soil microbial and biochemical changes associated with reduced tillage. *Soil Sci. Soc. Am. J*. 44:765-771.

Doran, J.W. 1987. Microbial biomass and mineralizable nitrogen distributions in no-tillage and plowed soils. *Biol. Fert. Soils* 5:68-75.

Doran, J.W. and M.S. Smith. 1987. Organic matter management and utilization of soil and fertilizer nutrients. p. 53-72 In: R.F. Follett, J.W.B. Stewart, and C.V. Cole (eds.), *Soil fertility and organic matter as critical components of production systems*. SSSA special publication # 19. ASA, Madison, WI.

Doran, J.W. and M.S. Smith. 1991. Overview: Role of cover crops in nitrogen cycling. p. 85-90 In: W. L. Hargrove (ed.), Proceedings of an International conference *"Cover crops for clean water"*. Soil Conser. Soc. Amer., Ankeny, IA.

Doran, J.W. and M.R. Werner. 1990. Management and soil biology. p. 205-230 In: C.A. Francis, C.B. Flora, and L.D. King (eds.) *Sustainable agriculture in Temperate Zones*. John Wiley & Sons, New York.

Doran, J.W., D.G. Fraser, M.N. Culik, and W.C. Liebhardt. 1987a. Influence of alternative and conventional agricultural management on soil microbial processes and N avaiability. *Am. J. Alternative Agric*. 2:99-106.

Doran, J.W., L.N. Mielke, and J.F. Power. 1987b. Tillage/residue management interactions with the soil environment, organic matter, and nutrient cycling. *Intecol Bull*. 15:33-39.

Doupnik, B., Jr., M.G. Boosalis, G. Wicks, and D. Smika. 1975. Ecofallow reduces stalk rot in grain sorghum. *Phytopathol*. 65:1021-1022.

Dowdell, R.J. and R. Crees. 1980. The uptake of ^{15}N-labeled fertilizer by winter wheat and its immobilization in a clay soil after direct drilling or plowing. *J. Sci. Food Agric*. 31:992-996.

Duxbury, J.M., M.S. Smith, and J.W. Doran. 1989. Soil organic matter as a source and sink of plant nutrients. p. 33-67(Chapter 2) In: D. C. Coleman, J.M. Oades, and G. Uehara (eds.), *Dynamics of soil organic matter in tropical soils*. NIFTAL, University of Hawaii, Honolulu.

Elliott, L.F., V.L. Cochran, and R.I. Papendick. 1980. The effect of residue management on microbial root colonization, straw decomposition, and toxin production. p. 88-96. In: C.D. Fanning (ed.), *Proceedings, Tillage Symposium*. North Dakota State University, Fargo.

Follett, R.F. and D.S. Schimel. 1989. Effect of tillage practices on microbial biomass dynamics. *Soil Sci. Soc. Am. J.* 53: 1091-1096.

Fox, R.H. and V.A. Bandel. 1986. Nitrogen utilization with no-tillage. p. 117-148. In: M.A. Sprague and G.B. Triplett (eds.), *No-tillage and surface-tillage agriculture: The tillage revolution.* John Wiley & Sons, New York.

Greenland, D.J. 1975. Bringing the green revolution to the shifting cultivator: Better seed, fertilizers, zero or minimum tillage, and mixed cropping are necessary. *Science* 190:841-844.

Guenzi, W.D. and T.M. McCalla. 1962. Inhibition of germination and seedling development by crop residues. *Soil Sci. Soc. Am. Proc.* 26:456-458.

Guenzi, W.D. and T.M. McCalla. 1966. Phenolic acids in oats, wheat, sorghum, and corn residues and their phytotoxicity. *Agron. J.* 58:303-304.

Guenzi, W.D., T.M. McCalla, and F.A. Norstadt. 1967. Presence and persistence of phytotoxic substances in wheat, oat, corn, and sorghum residues. *Agron. J.* 59:163-165.

Hardin, B. 1989. Coping with crop residues. *Agricultural Res.* March: p. 19-20.

Harper, S.H.T. and J.M. Lynch. 1981. The kinetics of straw decomposition in relation to its potential to produce the phytotoxin acetic acid. *J. Soil Science* 32:627-637.

Harwood, R.R. 1990. A history of sustainable agriculture. p. 3-19 In: C.A. Edwards et al. (ed.), *Sustainable agricultural systems.* Soil and Water Conser. Soc. Am., Ankeny, IA.

Heritage, A.D., A.D. Rovira, G.D. Bowen, and R.L. Correll. 1989. Influence of soil water on the growth of *Gaeumannomyces graminis* var. *tritici* in soil: Use of a mathmatical model. *Soil Biol. Biochem.* 21:729-732.

House, G.J., B. J. Stinner, D.A. Crossley, Jr., E.P. Odum, and G.W. Langdale. 1984. Nitrogen cycling in conventional and no-tillage agroecosystems in the Southern Piedmont. *J. Soil Water Conser.* 39:194-200.

Huber, D.M. and R.D. Watson. 1974. Nitrogen form and plant disease. *Ann. Rev. Phytopathol.* 12:139-165.

Jenny, H. 1980. *The soil resource: origin and behavior.* Ecological studies 37. Springer-Verlag, New York.

Karlen, D.L. and J.W. Doran. 1991. Cover crop management effects on soybean and corn growth and nitrogen dynamics in an on-farm study. *Amer. J. Alternative Agric.* 6:71-81.

Kimber, R.W.L. 1967. Phytotoxicity from plant residues. I. The influence of rotted wheat straw on seedling growth. *Aust. J. Agric. Res.* 18:361-374.

Kimber, R.W.L. 1973. Phytotoxicity from plant residues. II. The effect of time of rotting of straw from grasses and legumes on the growth of wheat seedlings. *Plant Soil* 38:347-361.

Kitur, B.K., M.S. Smith, R.L. Blevins, and W.W. Frye. 1984. Fate of ^{15}N-depleted ammonium nitrate applied to no-tillage and conventional tillage corn. *Agron. J.* 76:240-242.

Liebhardt W.C., R.W. Andrews, M.N. Culik, R.R. Harwood, R.R. Janke, J.K. Radke, and S.L. Rieger-Schwartz. 1989. Crop production during conversion from conventional to low-input methods. *Agron. J.* 81:150-159.

Linderman, R.G. and T.A. Toussoun. 1968. Predisposition to *Thielaviopsis* root rot of cotton by phytotoxins from decomposing barley residues. *Phytopathol.* 58:1571-1574.

Linn, D.M. and J.W. Doran. 1984a. Aerobic and anaerobic microbial populations in no-till and plowed soils. *Soil Sci. Soc. Am. J.* 48:794-799.

Linn, D.M. and J.W. Doran. 1984b. Effect of water-filled pore space on CO_2 and N_2O production in tilled and nontilled soils. *Soil Sci. Soc. Am. J.* 48:1267-1272.

Logan, T.J., J.M. Davidson, J.L. Baker, and M.R. Overcash. 1987. *Effects of Conservation tillage on groundwater quality: nitrates and pesticides.* Lewis Publishers Inc., Chelsea, MI.

Lynch, J.M. 1977. Phytotoxicity of acetic acid produced in the anaerobic decomposition of wheat straw. *J. Appl. Bacteriol.* 42:81-87.

Lynch, J.M. 1978. Production and phytotoxicity of acetic acid in anaerobic soils containing plant residues. *Soil Biol. Biochem.* 10:131-135.

Lynch, J.M. 1987. Allelopathy involving microorganisms. p. 44-52. In: G.R. Waller (ed.), *Allelochemicals: Role in agriculture and forestry.* Am. Chem. Soc. Sym. Series 330. Am. Chem. Soc., Washington, D.C.

Lynch, J.M. and L.F. Elliott. 1983. Minimizing the potential phytotoxicity of wheat straw by microbial degradation. *Soil Biol. Biochem.* 15:221-222.

Lynch, J.M. and L.M. Panting. 1980. Cultivation and the soil biomass. *Soil Biol. Biochem.* 12:29-33.

Magdoff, F.R., D. Ross, and J. Amadon. 1984. A soil test for nitrogen availability in corn. *Soil Sci. Soc. Am. J.* 48:1301-1304.

McCalla, T.M. and T.J. Army. 1961. Stubble mulch farming. *Adv. Agron.* 13:125-196.

McCalla, T.M. and F.L. Duley. 1948. Stubble mulch studies: effect of sweetclover extract on corn germination. *Science* 108:163.

McCalla, T.M. and F.A. Haskins. 1964. Phytotoxic substances from soil microorganisms and crop residues. *Bact. Rev.* 28:181-207.

McCalla, T.M. and F.A. Norstadt. 1974. Toxicity problems in mulch tillage. *Agric. and Envrion.* 1:153-174.

McCalla, T.M., W.D. Guenzi, and F.A. Norstadt. 1963. Microbial studies of phytotoxic substances in the stubble-mulch system. *Z. Allg. Mikrobiol.* 3:202-210.

McDonald, A. 1941. *Early American soil conservationists.* Misc. Publ. No. 449. U.S. Dept. Agric., Soil Conserv. Service. Reprinted Oct. 1959. U.S. Gov't Printing Office, Washington, D.C.

Mengel, D.B., D.W. Nelson, and D.M. Huber. 1982. Placement of nitrogen fertilizers for no-till and conventional till corn. *Agron. J.* 74:515-518.

Menzies, J.D. and R.G. Gilbert. 1967. Responses of the soil microflora to volatile components in plant residues. *Soil Sci. Soc. Am. Proc.* 31:495-496.

Mielke, L. N., J.W. Doran, and K.A. Richards. 1986. Physical environment near the surface of plowed and no-tilled surface soils. *Soil Tillage Res.* 5: 355-366.

Mowitz, D. 1991. Fertilizing options. *Successful Farming.* May/June: p. 24.

National Research Council. 1989. *Alternative Agriculture: Committee on the role of alternative farming methods in modern production agriculture.* Board on Agriculture, National Research Council. National Academy Press, Washington, D.C.

Norstadt, F.A. and T.M. McCalla. 1963. A phytotoxic substance from a species of *Penicillium. Science* 140:410-411.

Norstadt, F.A. and T.M. McCalla. 1969. Patulin production by *Penicillium urticae* Bainier in batch culture. *Appl. Microbiol.* 17:193-196.

Patrick, Z.A. 1971. Phytotoxic substances associated with the decomposition in soil of plant residues. *Soil Sci.* 111:13-18.

Patrick, Z.A. and L.W. Koch. 1958. Inhibition of respiration, germination and growth by substances arising during the decomposition of certain plant residues in the soil. *Can. J. Bot.* 36:621-647.

Patrick, Z.A. and L.W. Koch. 1963. The adverse influence of phytotoxic substances from decomposing plant residues on resistance of tobacco to black root rot. *Can. J. Bot.* 41:747-758.

Pereira, H.C. 1975. Agricultural science and the traditions of tillage. *Outlook Agri.* 8:211-212.

Phillips, R.E. and S.H. Phillips. 1984. *No-tillage agriculture: Principals and Practices.* Van Nostrand Reinhold, New York.

Phillips, R.E., R.L. Blevins, G.W. Thomas, W.W. Frye, and S.H. Phillips. 1980. No-tillage Agriculture. *Science* 208:1108-1113.

Plant Protection Limited. 1973. Minimum tillage issue. *Outlook Agri.* 7:142-200.

Plant Protection Limited. 1975. Reduced cultivation and direct drilling issue. *Outlook Agri.* 8:211-260.

Power, J.F. 1987. *The role of legumes in conservation tillage systems.* Proceedings of a national conference, University of Georgia, Athens, April 27-29, 1987. Soil Conserv. Soc. Am.. Ankeny, IA.

Power, J.F. and J.O. Legg. 1978. Effect of crop residues on the soil chemical environment and nutrient availability. p. 85-100. In: W.R. Oschwald (ed.), *Crop Residue Management Systems.* Am. Soc. Agron. Spec. Pub. 31. Madison, WI.

Power, J.F., and R.I. Papendick. 1985. Organic sources of nutrients. p. 503-520. In: O.P. Engelstad (ed.), *Fertilizer Technology and Use, Third Edition.* Soil Sci. Soc. Am., Madison, WI.

Power, J.F., J.W. Doran, and W.W. Wilhelm. 1986. Uptake of nitrogen from soil, fertilizer, and crop residues by no-till corn and soybean. *Soil Sci. Soc. Am. J.* 50:137-142.

Putnam, A.R. and J. DeFrank. 1983. Use of phytotoxic plant residues for selective weed control. *Crop Protection* 2:173-181.

Rice, E.L. 1984. *Allelopathy, 2nd edition.* Academic Press, Inc., New York. 422 p.

Rice, C.W. and M.S. Smith. 1982. Denitrification in no-till and plowed soils. *Soil Sci. Soc. Am. J.* 46:1168-1172.

Rice, C.W. and M.S. Smith. 1984. Short-term immobilization of fertilizer nitrogen at the surface of no-till and plowed soils. *Soil Sci. Soc. Am. J.* 48:295-297.

Rice, C.W., M.S. Smith, and R.L. Blevins. 1986. Soil nitrogen availability after long-term continuous no-tillage and conventional tillage corn production. *Soil Sci. Soc. Am. J.* 50:1206-1210.

Rothrock, C.S. 1987. Take-all of wheat as affected by tillage and wheat-soybean doublecropping. *Soil Biol. Biochem.* 19:307-311.

Rothrock, C.S. and B.M. Cunfer. 1986. Absence of take-all decline in double-cropped fields. *Soil Biol. Biochem.* 18:113-114.

Sarrantonio, M. and T.W. Scott. 1988. Tillage effects on availability of nitrogen to corn following a winter green manure crop. *Soil Sci. Soc. Am. J.* 52: 1661-1668.

Schertz, D.L. 1988. Conservation tillage: An analysis of acreage projections in the United States. *J. Soil Water Conserv.* 43:256-258.

Smika, D.E. and G.A. Wicks. 1968. Soil water storage during fallow in the Central Great Plains as influenced by tillage and herbicide treatments. *Soil Sci. Soc. Am. Proc.* 32:591-595.

Smith, A.M. 1973. Ethylene as a cause of soil fungistatsis. *Nature* 246:311-313.

Smith, A.M. 1976. Ethylene production by bacteria in reduced microsites in soil and some implications to agriculture. *Soil Biol. Biochem.* 8:293-298.

Soil Conservation Society of America. 1973. *Conservation tillage: the proceedings of a national conference.* Held March 28-30, 1973 in Des Moines, IA. Soil Conservation Society of America. Ankeny, IA.

Soil Conservation Society of America. 1983. Special Issue: Conservation tillage. *J. Soil Water Conserv.* 38:134-319.

Sprague, M.A. and G.B. Triplett. 1986. *No-tillage and surface tillage agriculture: The tillage revolution.* John Wiley & Sons, New York.

Sumner, D.R., B. Doupnik, Jr., and M.G. Boosalis. 1981. Effects of reduced tillage and multiple cropping on plant diseases. *Ann. Rev. Phytopathol.* 19:167-187.

Tang, C-S. and A.C. Waiss. 1978. Short-chain fatty acids as growth inhibitors in decomposing wheat straw. *J. Chem. Ecology* 4:225-232.

Timmons, D.R., J.L. Baker, and R. Cruse. 1987. Corn growth in a no-till system as influenced by method, placement, and time of nutrient application. p. 47-52. In: *Proceedings of the fluid fertilizer symposium.* March 16-18, 1987, Clearwater Beach, FL.

Touchton, J.T. and W.L. Hargrove. 1982. Nitrogen sources and methods of application for no-tillage corn production. *Agron. J.* 74:823-826.

Toussoun, T.A. and Z.A. Patrick. 1963. Effect of phytotoxic substances from decomposing plant residues on root rot of bean. *Phytopathol.* 53:265-270.

Toussoun, T.A., A.R. Weinhold, R.G. Linderman, and Z.A. Patrick. 1968. Nature of phytotoxic substances produced during plant residue decomposition in soil. *Phytopathol.* 58:41-45.

Triplett, G.B., Jr. 1982. Tillage and crop productivity. p. 251-262. In: M. Rechcigl, Jr. (ed.), *CRC Handbook of Agricultural Productivity, Vol. 1, Plant Productivity.* CRC Press, Inc., Boca Raton, FL.

U.S. Department of Agriculture. 1975. *Minimum tillage: A preliminary technology assessment.* W.B. Back (assessment leader). Office of Planning and Evaluation, USDA, Washington, D.C.

Unger, P.W. and T.M. McCalla. 1980. Conservation tillage systems. *Adv. Agron.* 33:1-58.

Wagger, M.G. 1989. Time of desiccation effects on plant composition and subsequent nitrogen release from several winter annual cover crops. *Agron. J.* 81:236-241.

Wanzel, R.J. 1988. Point injector rolls out. *Solutions* May/June: p. 14-15.

Woodruff, N.P., C.R. Fenster, W.H. Harris, and M. Lundquist. 1966. Stubble-mulch tillage and planting in crop residues in the Great Plains. *ASAE Trans.* 9:849-853.

Dynamic Processes of Vesicular-Arbuscular Mycorrhizae: A Mycorrhizosystem within the Agroecosystem

S.F. Wright and P.D. Millner

This chapter was prepared by U.S. government employees as part
of their official duties and legally cannot be copyrighted.

I. Introduction

In the ecosystem concept, agroecosystems are characterized by major dependence on and influence by factors external to the system (Odum, 1984), such as
energy and agricultural chemicals and their residues. Unlike natural solar-
powered ecosystems, agroecosystems are created and controlled by human
management of ecological processes for production, productivity, and conservation. The challenge confronting the agricultural community is to reduce the input
and output costs to the agricultural system so that these costs are integrated
compatibly at the farming scale. Utilization of more internal control mechanisms
that can function as low-energy subsystem feedbacks with high-energy effects
like those of natural ecosystems (see Patten and Odum, 1981) will enhance the
self-sustaining capacity of agroecosystems. The soil module of agroecosystems
is a prime arena for enhancement of internal control mechanisms because it is
a biologically dynamic, rather than static resource. Within this soil module, the
rhizosphere is the locus of greatest rate of flow of energy and minerals among
the biological, physical, and chemical components; it can be considered as a
subsystem. Vesicular-arbuscular mycorrhizal (VAM) fungi and the mutualistic
associations they form with plants are also subsystems which perform a unique
and important functional role in the rhizosphere of most agroecosystems.

Hyphae of VAM fungi serve as conduits to plants of nutrients such as
phosphorus, microelements, nitrogen and water by bridging all areas of the
rhizosphere, i.e. endorhizosphere, rhizoplane and ectorhizosphere (Lynch,
1982). Outflow from hyphae to areas beyond the influence of root exudates
provides carbon for the heterotrophic soil biota. Soil structure and organic
matter turnover are affected by increased biotic activity in the area explored by
VAM fungi.

A unifying feature of the interactions across ecosystem scales is the
continuous cycling between biotic and abiotic components within and among the
defined systems from global to VAM associations (Figure 1). At all levels,
activities of closely associated biological components interdependently influence
nutrient cycling and energy flow to different tropic levels within the ever-
changing physical environment. Unique among microorganisms, the root-like
functions of VAM fungi establish a rhizosphere within the root rhizosphere
[mycorrhizosphere according to Linderman (1988)]. We would like to go one
step further and refer to the VAM subsystem within the rhizosphere as the
mycorrhizosystem. This subsystem is characterized by the type of self-sustained
cycling and internal controls that can enhance the efficiency of a managed
ecosystem such as an agroecosystem.

Scaling the "levels of organization" of biological systems from the agroecosystem to the rhizosphere and VAM subsystems (Figure 1) in accordance with
Hierarchy Theory (O'Neill et al., 1991), we identify and describe constraints by
lower-level components in terms of direct and indirect inputs, transformations,
and outputs of the plant-fungus association. Certain assumptions are made to
assist in determining the important components to be included in the mycorrhizo-

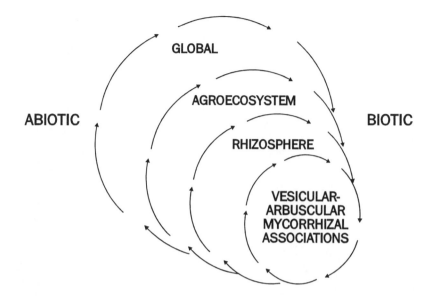

Figure 1. A conceptual integration of scales of cycles within the biosphere. From global down to the VAM fungal symbiotic association cycling among and between abiotic and biotic components is occurring.

system model. Citations are representative, not all-inclusive, of the topics discussed. Influences of the rhizosphere which affect colonization and extra-matrical hyphae are considered, but the reader is referred to Gianinazzi (1991) for discussion of the biochemical interactions between fungi and plant cells.

The mycorrhizosystem is characterized by dynamic processes affecting VAM interactions with other soil biota and physical and chemical constituents in the rhizosphere over time (Figure 2). Interactions among the separate components occur in a matrix where water, temperature, and soil physical characteristics influence the amount of activity. Climate is reflected in this diagram by soil temperature and the amount of water and sunlight plants receive. Thirty years of work on VAM fungi has elucidated much about VAM fungi and their function. This review is a discussion of how the separate components identified in Figure 2 fit into the functional mycorrhizosystem. We discuss facts which have been revealed about the separate components, but we take the position that the measure of functional mycorrhizosystems should be a focus of more effort for two reasons: (1) to support low-input agriculture, and (2) because indepen-dent control of most factors in Figure 2 is not feasible.

Characteristics of the mycorrhizosystem (Figure 2) are introduced here to put into perspective the complexities of agroecosystem processes and the mycor-rhizosystem (Section II). In Figure 2, components of the mycorrhizosystem are grouped according to whether the *primary* effect is on growth and activities of

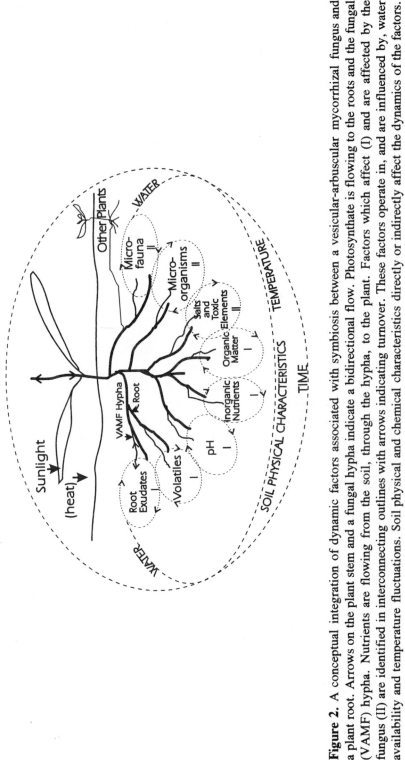

Figure 2. A conceptual integration of dynamic factors associated with symbiosis between a vesicular-arbuscular mycorrhizal fungus and a plant root. Arrows on the plant stem and a fungal hypha indicate a bidirectional flow. Photosynthate is flowing to the roots and the fungal (VAMF) hypha. Nutrients are flowing from the soil, through the hypha, to the plant. Factors which affect (I) and are affected by the fungus (II) are identified in interconnecting outlines with arrows indicating turnover. These factors operate in, and are influenced by, water availability and temperature fluctuations. Soil physical and chemical characteristics directly or indirectly affect the dynamics of the factors.

VAM fungi (Group I) or the component is affected by the presence and activity of VAM fungi (Group II). This classification is a simplification of component-VAM fungi and component-component interactions, but it is hoped that this illustration will be useful to emphasize the complex activities of the mycorrhizosystem. Individual components may cycle through low to high influence on the system inherently more rapidly than others, or moisture and temperature may intermittently drive one or several components at a rapid rate. We cannot use terms such as normal or average to describe rates of processes because currently such isolated values are not definable.

In Group I root exudates impact significantly on VAM fungal life cycle. Volatiles such as CO_2, O_2, and organics from roots are in continuous flux and influence germination of spores and growth of hyphae. Soil pH is the chemical property most often related to the effectiveness of isolates of VAM fungi. The primary effects of mineral nutrients in relation to VAM fungi are on plant nutrient status which, in turn, affects VAM fungi. Yet to be determined is whether calcium plays a role in VAMF nutrition aside from the change in soil pH brought about by calcium salts. Organic matter is important in relation to most of the above-mentioned influences on VAM fungi, i.e. volatiles, pH, cycling of nutrients, and moisture.

Components in Figure 2 which are affected by the presence or activity of VAM fungi are in Group II. Qualitative and quantitative changes in root exudates brought about by VAM colonization and transport of energy compounds via VAM fungal hyphae influence the activity of microbial processes. Toxic elements such as manganese cycle between reduced and oxidized states by the activity of microorganisms affected by exudates of VA mycorrhizal roots. Micro-fauna populations, which include fungus-grazers such as nematodes and collembola, fluctuate as hyphae become available.

II. Experimental Constraints

This discussion focuses on limitations of technology and interpretation of results. Suggestions are made for improvements where possible.

A. Sensitivity of Measurements

Investigative methods and techniques used to date generally have relied upon point-in-time measurements or have been inadequate to accurately measure chemical and biological changes at the level of sensitivity required to make definitive assessments of dynamic changes. Mini-rhizotrons and other *in situ* visualization techniques are not sensitive enough at the current stages of development to detect interactions between roots and VAM fungi. Chambers such as the one used by Friese and Allen (1991) may prove very useful for studies of biological and chemical interactions with specific VAM fungi.

"Soilscopes" such as those used in medical procedures (orthoscopes or hysteroscopes) may be useful tools but high magnification would be needed, and the soil-root-fungus environment may be disturbed to the extent that observations would not reflect natural conditions. Measurement of small and transient chemical changes present another challenge. Current state-of-the art instrumentation makes identification of chemical constituents possible, but measurements of small changes in short time periods in a field soil remain impossible.

B. Diversity of Fungi and Experimental Conditions

Generally, investigators use isolates of one to a few species. Also, experimental conditions often are limited although they range from highly artificial *in vitro* chambers to field soils. Often results from individual experiments are generalized to represent behavior of all VAM fungi. This does not take into account that plant species, plant densities, soils, microflora, VAM fungal populations, and fungal propagules exert profound influences on the association. For example, the type of propagule and its maturity may affect the rate of colonization and the cascade of events that follow. Because of this lack of uniformity of experimental conditions, we caution that there are considerable gaps in knowledge about many aspects of the life cycle and function of these fungi under field conditions.

By using isolates from the same low pH soil, differences in plant responses among naturalized VAM fungi can be shown. Comparisons of plant growth, tissue P concentration, and mycorrhizal infection of roots indicated that an isolate of *Acaulospora laevis* was more effective than an isolate of *Gigaspora margarita* on serecia lespediza and Ladino white clover (Wilson 1988). Also, differences in effectiveness among isolates of the same species have been shown (Bethlenfalvay et al., 1989) by using *Gl. mosseae* isolates from various geographical regions to inoculate soybean. Significant differences in plant dry mass, leaf K, N and P concentrations along with other measurements of plant response were found.

Fungal diversity has been determined from morphological characters of the spores produced by mycorrhizae (Morton and Benney, 1990). In addition, taxonomic groups can be delineated by morphological characters linked to developmental patterns of internal fungal structures (Abbott, 1982), fungal structures but it is uncertain that these characters or spore morphology are linked to functional diversity. There has been such an emphasis on delineation of species that any plant and soil response that may be common to higher taxonomic entities (genera, families) have not been discovered.

The species concept remains unclear despite recent efforts (Morton, 1990b) to elucidate it more clearly. VAM species are not unified entities and, therefore, cannot be treated as single experimental units (Morton, personal communication). A serious semantic problem exists in the literature which leaves an erroneous impression as to which biological entities are actively participating in mycorrhizal interactions. For example, "species A" is reported as producing a certain result, when the investigator really means an isolate of "species A". The experimental isolate originating from a given site and propagated in pot culture is genetically isolated from all other individuals of the species, and may share important morphological, but not physiological properties with isolates at disparate locations. Until work is focused on the proper level of comparison in experimental systems (between isolates and populations, rather than between species) real genetic and phenotypic differences critical to the functioning of the association will not be recognized.

Investigators who want to support comparisons at the isolate level are strongly encouraged to assign and report isolate numbers and to deposit cultures and voucher specimens for future reference in a culture collection such as INVAM (International Culture Collection of VA Mycorrhizal Fungi; Joseph Morton, Curator; West Virginia University; Morgantown, WV 26506-6057).

At global and agroecosystem scales, the interactions of soil parent material, water, and air have led to large areas where soil moisture, texture, fertility and the micro- and macro-fauna associated with the fertility status, are favorable for establishment of VAM symbiosis. Specific combinations of biotic and abiotic conditions have led to the evolution of small changes in the symbiosis which are reflected in adaptations of fungal isolates and hosts at each site (Morton, 1990a).

C. Time

The time course and complexity of the interaction between VAM fungi and roots afford many opportunities for the association to fail from the standpoint of achieving symbiosis. Research results utilizing plants are generally assessed by the success or failure of establishment and function of the symbiosis as reflected in enhanced mineral uptake. Plants integrate the complex biological and physiological interactions between roots and VAM fungi. Influences on the life cycle of VAM fungi have been reviewed (Bowen, 1987; Hetrick, 1984). Spore maturation, spore germination, germ tube growth, interception of a susceptible root, an interactive association between root cells and hyphae, and exchange of nutrients and biochemicals between fungus and plant over time are impacted by chemical, physical, and biological components in the rhizosphere. Some aspects of the association, such as arbuscular life span or time for infection to occur, have been investigated (see below). However, most rhizosphere changes are integrated into plant responses over weeks or months because of the necessity for development of the symbiosis over such a long time span. Therefore, we can comment that most aspects of time for specific rhizosphere changes, as related to VAM fungi, are unknown.

III. Agroecosystem Processes and the Mycorrhizosystem

Within the agroecosystem, in general terms, productivity can be assessed on a short or long term basis. The short term emphasizes primarily plant productivity with inputs delivered in such a manner as to uncouple dependence of plant productivity from soil biology. In contrast, the long term emphasizes soil-plant productivity which encompasses the concept of maintaining soil quality via microbial interactions to sustain production. VAM fungi and VAM fungal biomass are integral parts of soil-plant productivity because of their roles in (a) amelioration of environmentally induced plant stress, (b) soil structure development, and (c) carbon, nitrogen, and phosphorus cycling. These cycles involving VAM have been described and depicted by Allen (1991), Barea (1991), Bolan (1991), and Coleman et al. (1983), and the reader is referred to these for diagrammatic representations and specific details.

The mycorrhizal impact on all these processes is uniformly dependent on the development of an extramatrical hyphal network and its capacity to absorb, translocate, and transfer the respective nutrients to and from the host. Unlike other rhizosphere microorganisms, VAM fungi (at least at the species level) apparently maintain their niche despite considerable fluctuations in environmental conditions because of their obligate dependence on the host and their overall lack of host specificity. The patchy distribution of currently defined species occurs at all scales, from rhizosphere (Newman and Bowen, 1974) to landscape (Allen, 1991). This distribution of species reflects the vegetative and reproductive success of only a few of >150 VAM fungal species confronted with the heterogenous soil-plant environment. Generalizations about the functions of VAM in nutrient cycles within the rhizosphere and ecosystem are based on experiments involving relatively few VAM fungal species with few intraspecific (isolate) comparisons. Characteristics that affect the functioning of VAM are now understood to vary significantly at the species and isolate level, e.g. rate and amount of hyphal network development. Future research needs to address the overall contribution of VAM to the rates, amounts, and forms of carbon, nitrogen, and phosphorus cycling through the rhizosphere with particular attention on soil productivity along with plant productivity in the long term perspective. Sub-system models that can iteratively project impacts of influences such as grazing, drought, tillage or temperature would be immensely valuable for projecting short and long term consequences of rate and quantitative or qualitative changes in VAM. They are especially needed to help researchers identify critical information/data gaps. With such a basis, answers to the following can be approached more quantitatively. How do changes at the ecosystem level affect VAM fungi and VAM functioning? How do changes in VAM fungi, in turn, affect, mycorrhizosystem and ecosystem processes in the short and long term?

At the present stage, investigators must still assess the impact of particular disturbances, for example tillage, on specific plant-fungus-soil combinations. Effort is needed to acquire information on system level impacts of disturbances, with consideration given to variability in VAM fungal isolate effectiveness (see below Section III, B).

The influence of VAM on carbon cycling occurs through the allocation of host carbon to roots and (a) its translocation to mycorrhizal hyphae which (b) respire it to CO_2 and (c) incorporate it into hyphal biomass, and (d) exude it in the form of carbohydrates to soil heterotrophs. In addition, VAM fungi are also involved in the decomposition of organic carbon (Allen and MacMahon, 1985; Warner, 1984) at localized sites in soil through the proliferation and activity of extraradical hyphae (St. John et al., 1983). Because VAM fungal hyphae ramify through the microsites and pores of soil, they change the spatial distribution of carbonaceous compounds. In general, fungi are the basis for soil food webs under no tillage agroecosystems (Hendrix et al., 1986), and VAM fungal hyphae serve as substrates for fungus-grazing insects (Rabatin and Stinner, 1985).

Jakobsen and Rosendahl (1990) reported that below-ground respiration of mycorrhizal and nonmycorrhizal plants accounted for 27% and 12% respectively of the photoassimilated ^{14}C. Below-ground carbon allocation was 43% and 20% for mycorrhizal and nonmycorrhizal cucumbers colonized by an isolate of *G. fasciculatum*. They estimated C allocation to external hyphae at 1.5 mg d^{-1} i.e. ca. 4% of fixed ^{14}C. By assuming an equivalent allocation of C to internal and external hyphae, they estimated C allocation to internal hyphae at 5.9 mg d^{-1} or 16% of fixed CO_2 and hence an overall allocation to VAM respiration and biomass of 20% of fixed C.

Mycorrhizal influence on the N cycle occurs by immobilization of N in hyphal and spore biomass, uptake and translocation of N from microsites in the soil to host plants and between host plants (Ames et al., 1983; Hamel et al., 1991; Van Kessel et al., 1985), transformation of inorganic N via nitrate reductase (Oliver et al., 1983), and indirect enhancement of N-fixation (Barea and Azcón-Aguilar, 1983).

Mycorrhizal influence on the P cycle involves not only the hyphal uptake, translocation, and transfer of P from soil to plant (Smith and Gianinazzi-Pearson, 1988), but also the solubilization of unavailable sources of P and the faster rate of absorption of soil P to plant (see Bolan, 1991). Rates of inflow of P to mycorrhizal roots are greater (17×10^{-14} moles cm^{-1} s^{-1}) than that to nonmycorrhizal roots (3.6×10^{-14} moles cm^{-1} s^{-1}) according to Sanders and Tinker (1973). Cox and Tinker (1976) reported the P flux across the arbuscular interface as 1.3×10^{-14} moles cm^{-2} s^{-1}.

A. Nonnutritional Physicochemical Influences

Management of agroecosystems often involves inputs which have a negative impact on mycorrhizosystem function. Recovery from such negative impacts will depend on the interaction of many factors (Figure 2). Given a "window of opportunity" such as a change in the crop or the tillage system, the inherent differences among VAM fungi in terms of survival, propagative, dispersal and plant-beneficial traits will drive the recovery process. The quantitative database on these fungal traits, even among isolates identified as "superior" in terms of their benefit to plant growth, is patchy and insufficient for modelling.

1. Grazing/Harvest/Fallow

Lack of suitable host plants (Ocampo and Hayman, 1981) and/or long-term fallowing (Thompson, 1987) can reduce the indigenous population of VAM fungi so drastically that growth of subsequent crops is severely negatively impacted. Heavy grazing of grasslands, considered the equivalent to harvesting in agronomic terms, imposes severe photosynthate stress in a semi-arid region and reduces VAM development significantly. (Bethlenfalvay et al., 1988).

2. Soil Fumigation

In agroecosystems, use of soil fumigants and fungicides has the immediate benefit of reducing populations of growth and pathogenic effects of targeted plant pathogens. However, nontargeted microbes in the mycorrhizosystem, especially VAM fungi, are reduced in numbers or in timeliness of growth and plant colonization (Menge, 1982). These specific disadvantages of fumigants and fungicides are eventually overcome with time and sometimes by virtue of additional input to the system via inoculation with VAM fungi. The cost of the short term gain has yet to be assessed in terms of the loss to the system in efficiency and cost of amelioration.

3. Water

The movement of water from soil to roots through the VAM hyphal pathway has been the subject of considerable experimentation (Nelsen, 1987). The production of copious hyphae by some isolates could have adaptive advantage in arid environments or environments subject to extremes of soil water availability. Hyphae are thought to access water held in soil aggregates. Controversy about VAM uptake of water is based on the confounding effect of the improved P status of mycorrhizal plants. However, Bethlenfalvay et al. (1988) have shown that growth of plants subjected to severe water stress is attributable to VAM-mediated increase in water uptake which is associated with increased soil hyphal biomass. Plant P status was equivalent for mycorrhizal and nonmycorrhizal plants. Other studies have shown that mycorrhizal *Pelargonium* (Sweatt and Davies, 1984), *Citrus* (Levy et al., 1983), and *Trifolium* (Safir et al., 1972) respond to water stress more rapidly than do nonmycorrhizal controls, but they recover more efficiently from water deficits.

4. Salt

The influence of chloride ions on VAM fungi is not clearly understood. Typical VAM formation with spore production has occurred in soils with >5000 ppm Cl⁻ (Bowen, 1980), whereas spore germination of an isolate of *Gigaspora margarita* is inhibited by the presence of chloride ion (Hirrel, 1981). More recently, Duke et al. (1986) showed that an isolate of *Glomus intraradices* enhanced salinity tolerance of a salt-sensitive citrus cultivar. In contrast, Graham and Syvertsen (1989) showed that there was no increase in salinity stress tolerance of two citrus cultivars when colonized by this same species. Physiological adaptation of fungi to increased Cl⁻ concentrations is known in general. The range of tolerance of VAM fungi and the quantitative effect on mycorrhizo-system outputs needs evaluation.

5. Temperature

Increased soil temperature stimulates VAM formation in onions (Furlan and Fortin, 1973), *Medicago* and *Trifolium* (Smith and Bowen, 1979), and stimulates spore germination of regionally adapted isolates (Schenck et al., 1975). Rates of translocation of P in mycorrhizae are sensitive to temperature changes (Cooper and Tinker, 1981). In general, all the biochemical transformation processes associated with VAM formation and function are potentially rate sensitive to temperature.

B. Soil Disturbance

System level disturbances of soil, such as tillage, impact major components in the rhizosphere, i.e. water (distribution and evaporation), temperature, and soil structure. Changes in these physical parameters in turn influence biological and chemical components of the rhizosphere qualitatively and quantitatively as shown in Figure 2. Mycorrhizae influence and are influenced by these microlevel biological and chemical components.

A major consequence of soil disturbance on mycorrhizae is the disruption of hyphal networks (Evans and Miller, 1988; Fairchild and Miller, 1988, 1990; Jasper et al., 1989, 1991) which are a considerable part of the inoculum available to field grown plants. Disturbance of these hyphae by tillage can lead to decreased root colonization by VAM fungi (Evans and Miller, 1988, 1990; Fairchild and Miller, 1988, 1990), and to decreased shoot dry mass and P concentration. In some cases, soil disturbance, even though severe, does not alter the amount of VAM fungal colonization, but still decreases shoot dry mass and P (McGonigle et al., 1990). The reasons for these contradictory results are not entirely clear, but several possible explanations can be offered. The extent or activity of the external mycelium may be greater in undisturbed soil systems, or existing networks might serve as sites for mycorrhizal extension when linked to new roots (McGonigle et al., 1990). Water, and consequently P, is more available in undisturbed field soil than in no-till soils. Inconsistencies in colonization results between different experiments may in part be due to differences in the mycorrhizal fungi that were present in the various experiments. In experiments involving field soils, even soils from the same field, colonization by different fungi even within a field may occur within the same or different growing cycles. Interpretations of results purported to show mycorrhizal causation of the disturbance effect, i.e. absence of effect after γ-irradiation and with nonmycorrhizal hosts, reduced effect after benomyl treatment (Evans and Miller, 1988), do not address the contributions of bacteria and nonmycorrhizal fungi on phosphorus solubilization in the overall effect.

Disruption of hyphal networks produced by mycorrhizal and nonmycorrhizal fungi essentially alters all microbial populations associated with the hyphae. The microbial populations change qualitatively and quantitatively in response to the type and rate of nutrient flow, root exudation, volatile generation, salt and toxic ion translocation, organic matter transformation, and pH. The new populations in turn exert their influence on the nutrient and elemental pools and transformation of complex molecules.

IV. Mycorrhizosphere Process Dynamics and VAM Fungi

The following section reviews research results pertaining to factors depicted in Figure 2. We reiterate that it is difficult to study one factor because interactions are complex, and that the results obtained for specific fungus:plant:soil interactions may not prove true for other combinations. It is particularly interesting to notice the results of experiments on specific microbes in field soils. Interactions among microbes or microbial by-products and plants occur so rapidly that experimental null hypotheses concerning the activity of a particular organism are often substantiated rather than refuted.

A. Rhizosphere Dimensions

The influence of the rhizosphere on infection has linear limits which can be estimated by the number of infection points and most probable number of infectious propagules. The apparent width of the rhizosphere of *Trifolium subterraneum* L. seedlings was defined as 2.5-6.5 mm for mixed soil inoculum and 8.9-13.2 mm for an isolate of *Gl. mosseae* (Smith et al., 1986). Root mass:vessel volume limited the reliability of data collected for the above study to 10- to 12-day-old seedlings. This type of experiment could be used to indicate rhizosphere influences on infection processes for different VAM fungal species for a variety of soils and plant genotypes, thereby broadening the knowledge base about variation in the VAM fungi(us):soil:plant genotype differences.

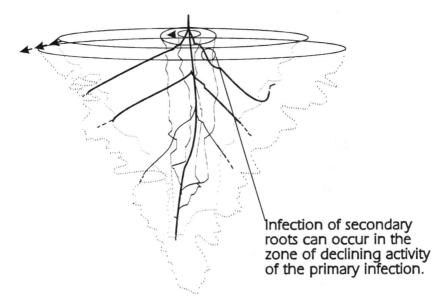

Infection of secondary
roots can occur in the
zone of declining activity
of the primary infection.

Figure 3. Infection and function of VAM fungi on a rhizosphere scale.
Primary plant root infection and function of the symbiosis is represented as
irregularly-shaped cones of increasing activity (outside) and decreasing
activity (inside). Irregularity of the outlines indicates micro-site perturbations of infection or function of the symbiosis.

We illustrate the dynamic movement of VAM fungal activity in the
rhizosphere in Figure 3. The main infection front and activity of mycelium in
part A are moving out and down from the primary root as the root grows. A
central core is decreasing in activity as arbuscules are senescing and the root
ages. Simultaneous with the activity occurring around primary roots, as lateral
roots emerge (B), a similar pattern of activity on a smaller scale is occurring.
This diagram is an integration of the major effects of root susceptibility to
infection, activity of arbuscules, and rhizosphere dynamics which will be
discussed below. Influences of factors represented in Figure 2 cause unevenness
in the flow of fronts of activity.

B. Root Exudates — Reducing Sugars and Organic Acids

The quantitative and qualitative influence of root exudates, as measured by total
sugar (Azcón and Ocampo, 1984) or total sugar and amino acids (Schwab et al.,
1983), and the relation to infection and spread have been included in the review
of the biology and physiology of infection (Bowen, 1987). There are changes in
root exudation with infection (see review by Schwab et al., 1991) and

differences between mycorrhizal host and nonhost plants (Schwab et al., 1984). Specific component(s) in the exudates which may influence infection are being discovered (i.e. isoflavonoids). Thus, we are left with exudates, measured as reducing sugars and amino acids, as a factor contributing to rhizosphere dynamics more from an influence on microflora than as critical factors in infection or spread.

C. CO_2 Combined with Root Exudates

The influences of CO_2 and root exudates on VAM fungal spores have been shown by the use of axenically grown transformed roots inoculated with an isolate of *Gi. margarita* (Bécard and Fortin, 1988; Bécard and Piché, 1989a, 1989b). An interaction between elevated CO_2 level and root exudates for growth of hyphae was demonstrated. The dynamics of this are interesting because of the synergy between CO_2 and root exudates required for fungal growth. Spores do not require nutrients or special conditions to germinate (Siqueira et al., 1985). Since elevated levels of CO_2 may be ephemeral in soils due to pockets of microbial activity, the synergism between root exudates and this gas is reasonable because a spore would not use its carbon reserves to elongate the germ tube unless a root were nearby.

Stimulated spore germination has been obtained in the presence of *Streptomyces* (Mugnier and Mosse, 1987) and certain bacteria (Mayo et al., 1986). This supports the idea of a microscale influence of CO_2 on spores.

D. Root Age and Architecture

Root susceptibility to infection over time is associated with rhizosphere width, and has implications for inoculum placement and for spread of infection. As discussed by Bowen (1987) there is still a need to experimentally determine the age of roots most susceptible to infection. More mechanistically, there is the need to determine the controlling factor(s) at the root-VAM fungal interface (Amijee, 1989; Schwab et al., 1991).

Adaptations in root morphology have been reviewed by Miller (1987) in relation to mycorrhizae. Coarseness of root morphology is loosely associated with mycorrhizal dependence, and plant species with highly developed root hairs often are nonmycorrhizal.

Hetrick et al. (1991) examined root architecture differences between warm- and cool-season grasses with and without the influence of mycorrhizae. Warm-season grasses respond to mycorrhizae by branching less frequently than noninfected roots, whereas cool-season grasses appear to have a fixed root architecture which is not altered by mycorrhizae. The authors speculate that conservation of energy expenditure by mycorrhizal warm-season grasses is involved.

E. Turnover of Roots

Root turnover is a major dynamic process in the mycorrhizosystem, yet it is often neglected simply because most observations are point-in-time assessments. The suggestion that VAM-infected roots may decompose faster than noninfected roots (Ames and Bethlenfalvay, 1987), indicates a need to investigate a different aspect of mycorrhizal impact on nutrient throughput via decomposition in addition to the typical transport functions.

F. Activity and Growth of Hyphae

Fungal biomass as well as measurements of actively transporting hyphae over time are critical in evaluating VAM fungal effectiveness. In a comparison of measurements of active (as determined by viable) hyphae by two different investigators (Schubert et al., 1987; Sylvia, 1988), Sylvia (1990) suggests that differences in turnover rates occur more rapidly under low-light than under high-light conditions. How does this tie into root turnover rate? Is it sufficient to measure hyphal turnover? Viable hyphae and actively transporting hyphae need to be differentiated. The chitin assay (Bethlenfalvay and Ames, 1987) may be useful for estimation of hyphal biomass, but there is still a need to quantify active hyphae. There is a recent report on the use of automated image analysis to determine viable fungal biomass in soils and on solid matrices (Morgan et al., 1991). There is a need to measure active, extramatrical hyphae of isolates of VAM fungi.

Hyphae are viable longer than they are functional in transport of nutrients as indicated by the ability of infected root fragments to contribute to infection potential of inoculants. Information on factors contributing to viability of hyphae needs further elucidation and quantification if possible. Although hyphae grow from cut ends of mycorrhizal roots (Hepper, 1984), most reports on hyphal growth-requirements and stimulators are from *in vitro* experiments using hyphae originating from spores.

Hepper and Warner (1983) show that organic matter per se has an important role in saprophytic growth of hyphae in soil. Cystine, glycine and lysine are specific requirements for hyphal growth from an isolate of *Gl. caledonius* (Hepper and Jakobsen, 1983). Whether these amino acids are required for hyphae to remain viable is not known. Information such as this on the requirements for saprophytic growth of a VAM fungus indicates limitation in the biosynthetic capabilities of this fungus.

Carr et al. (1985) reported stimulation of hyphal growth from chlamydospores of an isolate of *Gl. caledonius* by addition of wheat, lucerne and potato cells. The stimulating factor from lucerne cells was volatile. Cell extracts or exudates of *Pueraria phaseoloides* may be involved in stimulation of *Gi. margarita* hyphae (Paula and Siqueira, 1990).

A recent review by Strullu et al. (1991) summarizes their findings on regeneration of intraradical forms of *Glomus* spp. They describe an artificial medium for limited *in vitro* growth of intraradical hyphae and vesicles. Mycelium regenerated from vesicles and grown in pure culture does not lose its ability to infect roots, indicating that the medium supplies essential nutrients to maintain a saprophytic state at least for the short term.

Most current measurements are destructive; real time measurements of external hyphae in soil would be a major improvement. However, adaptation of a glass chamber such as that used by Friese and Allen (1991) for observing hyphal architecture, may provide a means for obtaining real-time measurements. Hyphae were classified as germ tubes, runner hyphae from adjacent infected roots, or hyphae from root fragments (Friese and Allen, 1991). Hyphae that grow out into the soil were characterized as runner hyphae or absorptive hyphae. Absorptive hyphae died back after 5-7 days and were not effective as units of infection. The type of *in vitro* observations obtained by Friese and Allen will be useful to further define rhizosphere influences on hyphae.

G. Arbuscule Life

Arbuscule life span contributes dramatically to the dynamics of nutrient transport. Alexander et al. (1989) determined microscopically that the active phase of the arbuscule is 2.5 days in six plant species colonized by an isolate of *Gl. fasciculatum*. Previous reports of arbuscule life span also indicate what is considered rapid decline of this active phase of the association (Cox and Tinker, 1976; Bevege and Bowen, 1975). Indexing the viability and activity of arbuscules with respect to mycorrhizal specific enzyme activity (Gianinazzi, 1991) among "superior" isolates would advance our knowledge of the structure/function relationship. The ratio of arbuscules per unit root length to extramatrical hyphal length might be considered for comparisons among fungal isolates with regard to functional transport capability. Technological advances that utilize automation need to be applied to this approach.

H. Inorganic Nutrients in the Rhizosphere

Changes in inorganic nutrients such as nitrogen in the rhizosphere may decrease formation and spread of infection. Addition of nitrate or ammonium to *T. subterraneum* was calculated by Smith and Walker (1981) to decrease the formation of new infections and the rate of fungal spread in the root cortex from each entry point. However, Thompson (1986) showed that nitrogen source (NH_4^+ or NO_3^-) in soilless cultures influenced VAM fungi primarily through modification of pH. Inhibition of infection was not seen at the concentration of N used in a typical full-strength nutrient solution (114 mg/l) for growth of maize or wheat roots colonized by isolates of *Gl. mosseae* or *Gl. fasciculatum*.

Phosphorus concentration in the plant, not in the rhizosphere, is correlated with VAM fungal infection and spread (Sanders, 1975). Localized placement of phosphate fertilizer by banding or topdressing does not affect the development of root infection in the fertilized zones differently from roots not in fertilized zones (Jasper et al., 1979). Sylvia and Neal (1990), using onion (*Allium cepa* L.) and isolates of *Gi. margarita* or *Gl. etunicatum* demonstrate an interaction between P and N. For a plant under nitrogen stress, additions of P did not affect root colonization by an isolate of *Gl. etunicatum*, but by doubling the N, there was a dramatic decrease in root colonization with added P.

Influences of other inorganic nutrients in the rhizosphere on VAM fungi need to be assessed. At the initiation of infection, the inorganic nutrient status of a soil is probably not a critical factor for the fungi. However, as plant growth and regrowth continue for one cycle or many, the native fertility status of the soil along with plant genotype, becomes important in the success of the symbiotic relationship.

I. Plant-Toxic Inorganic Elements and Salts in the Rhizosphere

A VAM fungus-mediated decrease in concentration of Mn in roots and shoots has been reported (McGee, 1987; Bethlenfalvay and Franson, 1989). Kothari et al. (1991) report a decrease in populations of Mn-reducing microorganism in the rhizosphere of plants infected with an isolate of *Gl. mosseae*. This is probably one of the clearest examples of changes in population associated with VAM fungal infection. The authors speculate that a selective decrease in a microbial population in the rhizosphere of an infected plant is mediated by changes in root exudation, by direct effects of fungal exudates, by stimulation of phytoalexin production in roots, or by decreased sloughing of cortical cells in basal root zones.

Selection of VAM fungal isolates tolerant of high concentrations of heavy metals (Gildon and Tinker, 1981) conditions may assist in bioremediation by metal accumulator plants. The role and impact of these organisms in heavy metal accumulation by plants must be accounted for in field situations.

J. Isoflavones and Phytoalexins

Isoflavonoid-stimulation of VAM fungal hyphal growth and root colonization *in vitro* studies has been shown (Nair et al., 1991). This class of compounds may be involved in the chemical dialog by which the symbionts recognize each other. The compounds are found in clover roots grown under P stress, and in a recent report Siqueira et al. (1991) demonstrate that addition of these compounds to inoculated soil increases root colonization.

In contrast with the stimulatory effects exerted by these compounds, phytoalexins and isoflavonoids that are inhibitory to phytopathogenic fungi have been isolated from mycorrhizal plants (Morandi et al., 1984; Morandi and Gianinazzi-Pearson, 1986). The presence of these resistance response compounds along with the formation of perisymbiotic membrane that delineates the arbuscular branches from the host cytoplasm, suggest that growth and spread of the fungal symbiont in the host is held in check. The susceptibility of individual VAM fungi to these resistance compounds could vary considerably in terms of their capacity to develop hyphal networks. There are no specific data on this but the technology is now available to make such determinations.

K. Chemotropism

Contact between germ tubes of a *Gi. gigantea* isolate and a root of a host plant leads to significant stimulation of root growth prior to colonization (Gemma and Koske, 1988). Also, emerging lateral roots respond by growing toward germinating spores of an isolate of *Gi. gigantea* (Gemma and Koske, 1988) and germ tubes of this fungus are attracted to bean roots through air to contact roots 11 mm away from the spore (Koske, 1982). The relative importance of this aspect of colonization will become clearer as the molecular signaling between host and fungus are elucidated.

L. Microorganisms and Microfauna

There are numerous positive and negative effects of soil microbes on the VAM life cycle (see Hetrick, 1986) and the sum total of these is what we need to focus on for determining the outcome of field situations. For this part of the mycorrhizosystem modelling effort, it will be nearly impossible and impractical to delve into the full range of interrelated responses quantitatively. For this reason we suggest that in order to predict mycorrhizosystem response to a range of agroecosystem management options, we need to assess the range of process outputs from the mycorrhizosystem. Possible approaches to quantification of process outputs include measurement and correlation of soil respiration, fungal biomass, and mycorrhizal-specific enzyme activity. Some microbial interaction that have been observed experimentally in the laboratory, but not necessarily at field scale, are described below to support our contention that mycorrhizosystem level processes need to be assessed.

1. Total Microbial Populations

The total microbial population at 0-5 mm from the root surface of mycorrhizal and non-mycorrhizal plants is similar (Meyer and Linderman, 1986; Kothari et al., 1991). This spatial zone is within the range of influenced proposed by Macfadyen (1969) for nutrient cycling activities at root-soil interfaces. Kothari et al. (1991) found that mycorrhizal plants had significantly higher total microbial populations at 5-15 mm and 15-25 mm from the root surface than non-mycorrhizal plants. The extensive hyphal growth at 25 mm from the root surface may have contributed substrate as hyphal exudates or decaying fungal structures.

Bacteria naturally associated with spores promote germination (Mayo et al., 1986). This has implications for *in vitro* experimentation under axenic conditions and provides another opportunity for study of signal molecules or indirect interactions between root exudates, soil microbes, and spores.

The tripartite symbiosis among legumes, rhizobia, and mycorrhizae is of interest because of the importance of these plants to agricultural systems. Azcón, et al. (1991) show different mycorrhizal species and/or *Rhizobium meliloti* treatments produce different effects on *Medicago sativa* growth. Bethlenfalvay et al. (1982) reported lower plant and nodule dry weight and nodule activity for bean (*Phaseolus vulgaris* L.) colonized by an isolate of *Gl. fasciculatum*. It is apparent that selection of effective mycorrhizal and rhizobial symbionts along with plant genotypes is necessary. Toward this end Bethlenfalvay et al. (1990) report on the use of the Diagnosis and Recommendation Integrated System (DRIS) to aid in this selection. The mycorrhizal impact on the N-fixation process is probably most important in low-input non-cash crop systems because effects on N-fixation per se are indirect.

2. Growth-stimulating Compounds of Microbial Origin

Bacteria produce grow-regulators which can affect roots directly or enhance the ability of VAM fungi to associate with roots. Reports of effects of bacterial supernatants on percent infection by an unidentified *Glomus* sp. (Azcón-G.and Barea, 1978) indicate the equivalence of these sources to pure plant hormones. There is always a question of numbers of bacteria which can bring about an effect equivalent to a known amount of hormonal stimulation. Soil microorganisms stimulate saprophytic development of an isolate of *Gl. mosseae* (Barea and Azcón-Aguilar, 1982) and thereby enhance infectivity of the fungus in controlled experiments using pasteurized soil (Azcón-Aguilar and Barea, 1985). Under field conditions, it is likely that hormonal effects are from limited *in situ* growth of a bacterial population which, in turn, is dependent upon encountering a pocket of nutrients.

Evidence that general activity of the soil biota affects VAM fungi has been obtained primarily by *in vitro* or axenic pot culture experiments. Germination rate of an isolate of *Gl. mosseae* is affected by the presence of free-living fungi (Azcón-Aguilar and Barea, 1986), and germination, hyphal growth and vegetative spore production are increased in the presence of some rhizosphere bacteria (Azcón, 1989). Inclusion of nonmycorrhizal microorganisms in experimental controls is now generally accepted as a requirement for proper interpretation of the effects on plant growth due to mycorrhizae as opposed to the effects due all other microbes (Ames and Bethlenfalvay, 1987).

3. Collembola and Nematodes

Grazing of the collembola *Folsomia candida* on VAM fungal hyphae was the probable cause of decreasing infection/unit of root length with increasing collembola density (Harris and Boerner, 1990). The plant response to increased densities of these fungal grazing animals was not, however, linear and the authors present scenarios for possible dynamics of the association of collembola with VAM fungi, other microorganisms, soil organic matter, and plants. For example, growth of plants may have been affected at low densities by collembola grazing on VAM and other fungi, but these animals may have affected an increase in available P through mineralization which offset the loss in uptake capacity from reduced VAM infection. Another explanation for greater P uptake, in spite of reduced colonization by VAM fungi, may have to do with increased metabolic activity of hyphae, thereby producing greater P uptake per unit of infected root. At higher grazing rates, the increase in P mineralization may have been insufficient to offset decreases in VAM infection. Feeding by microarthropods on hyphae and spores had been observed *in vitro* (Moore et al., 1985). *F. candida* fed on *Gl. fasciculatum* hyphae but not on hyphae of *Gl. mosseae* or *Gi. margarita*, while spores of *Gi. margarita* were attacked but not spores of *Gl. mosseae* or *Gl. fasciculatum*. Another observation made by Moore et al. (1985) was that hyphae were severed rather than entirely ingested which could impair nutrient transfer.

Effects of collembola on VAM fungi in axenic cultures can depend upon the stage of infection of plants at the time of inoculation with collembola. Collembola interferes with establishment of mycorrhizas in soybean (Kaiser and Lussenhop, 1991). However, in the same study, young mycorrhizal soybean roots slowed collembolan population growth.

VAM fungi and nematodes show interactions dependent upon the type of nematode, the species of VAM fungus, and the sequence of colonization by the fungus and exposure to nematodes (Bagyaraj, 1984). VAM fungal suppression of cyst nematodes in soybeans varied with time (Tylka et al., 1991). Use of VAM as a possible alternative to nematicides needs further investigation. Although reported studies do not indicate that nematodes are necessarily detrimental to spore production, our experience with pot cultures for spore production suggests that infestations by free-living, non-pathogenic nematodes severely decrease spore production in most cases.

4. Phosphate-Solubilizing Microorganisms

The possible synergistic interactions between VAM fungi and phosphate-solubilizing bacteria was proposed in 1975 by Barea et al. Interest in microorganisms which mobilize P has continued, particularly with respect to the use of such organisms as inoculants (Förster and Freier, 1988). Although solubilization of P by microorganisms has been shown *in vitro*, stimulation of plant growth by microbially produced hormones (Sattar and Gaur, 1987) may also be a factor which compounds the study of these organisms in relation to VAM fungi.

5. Interactions with Actinomycetes

It is a natural assumption that chitinase-producing actinomycetes would colonize VAM fungi because of the chitin content of fungal structures. It might also be assumed that such decomposers would be detrimental to growth of the organism that they attack. However, actinomycetes isolated from VAM fungal spores and reinoculated into nonsterile soil significantly increased hyphal densities and/or mycorrhizal root colonization by isolates of *Gl. macrocarpum* and *Gl. mosseae* on onion (Ames, 1989). Actinomycetes are probably involved in degradation and cycling of VAM fungal hyphae and spores over the long term.

V. Summary and Research Needs

This is an overview of the influence of VAM fungi on agroecosystem dynamics and the interactions of these fungi in more specific rhizosphere dynamics. Figure 4 summarizes our view of the controllable inputs to the agroecosystem which directly or indirectly influence the soil-plant-fungus interface. Management controls on the agroecosystem influence the rate and type of processing of nutrients etc. through the mycorrhizosystem. Optimal interfacing between the two systems results in efficient plant production, high soil productivity, and natural resource conservation.

What do we know about effects of management inputs on the mycorrhizo-system?

- Fallowing is detrimental to survival of propagules.
- Tillage disrupts hyphal networks to the detriment of the function of the mycorrhizosystem.
- Incorporating a nonmycorrhizal plant into a rotation is detrimental to survival of propagules.
- Edaphic factors such as pH, phosphorus status, or moisture inhibit colonization by certain isolates.
- Hyphal length in soil is an important determinant of mycorrhizal impact on plant performance.

Research needs for the near future should be directed at quantification of outputs of the mycorrhizosystem as influenced by changes in agroecosystem management. A constraint to measuring such outputs is the ability to quantitatively measure functional mycorrhiza. Gianinazzi et al. (1991) propose the use of phosphatase enzyme activity in a simple histochemical technique. Simple assays for quantification of functional hyphae in soil must be developed. While

S.F. Wright and P.D. Millner

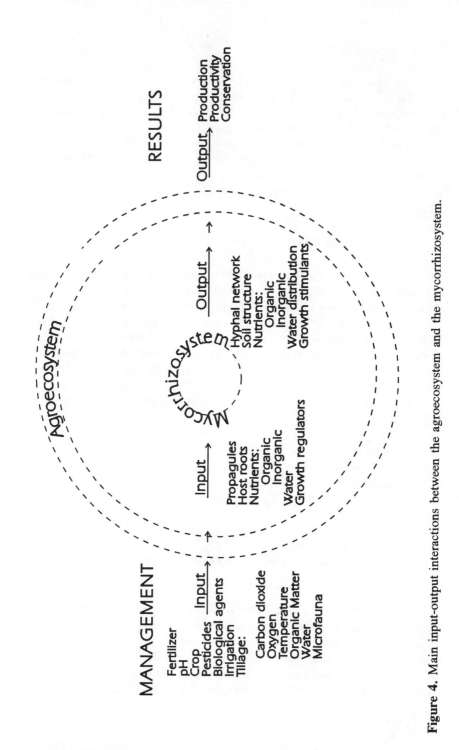

Figure 4. Main input-output interactions between the agroecosystem and the mycorrhizosystem.

tests for functional mycorrhizas and hyphae are being developed, field experiments should be continued. Demonstration of effects of soil disturbance on VAM hyphal network need to be extended from laboratory (Abbott and Robson, 1985; Evans and Miller, 1990) to field studies which include plant growth and nutrient status. Responses of populations of VAM fungi to changes in cropping are a factor in the agroecosystem approach which is being addressed (Johnson et al., 1991), but crop responses to changes in populations also need to be assessed.

We have attempted to identify some areas where more intense efforts on measurements need to be expended in order to quantify the effects of management practices on interfacing the mycorrhizosystem with the agroecosystem. Thirty years of observations can now be used to identify important information gaps and to help set priorities for further research. Conventional management for production as a primary goal has often resulted in uncoupling the mycorrhizosystem from the agroecosystem. Movement toward sustainable agriculture involves reestablishing and optimizing the interactions between these two systems.

Acknowledgments

The helpful discussions with D. M. H. Watson, J. B. Morton, and D. G. Kitt in the preparation of this manuscript are acknowledged. We thank T. A. Day and J. R. Ellis for reviewing the manuscript.

References

Abbott, L.K. 1982. Comparative anatomy of vesicular-arbuscular mycorrhizas formed on subterranean clover. *Aust. J. Bot.* 30:485-499.

Abbott, L.K. and A.D. Robson. 1985. Formation of external hyphae in soil by four species of vesicular-arbuscular mycorrhizal fungi. *New Phytol.* 99:245-255.

Alexander, T., R. Meier, R. Toth, and H.C. Weber. 1989. Dynamics of arbuscule development and degeneration in mycorrhizas of *Triticum aestivum* L. and *Avena sativa* L. with reference to *Zea mays* L. *New Phytol.* 110:363-370.

Allen, M.F. 1991. *The ecology of mycorrhizae.* Cambridge Univ. Press, New York, 184 pp.

Allen, M.F. and J.A. MacMahon. 1985. Impact of disturbance on cold dessert fungi: Comparative microscale dispersion patterns. *Pedobiologia* 28:215-224.

Ames, R.N. 1989. Mycorrhiza development in onion in response to inoculation with chitin-decomposing actinomycetes. *New Phytol.* 112:423-427.

Ames, R.N. and G.J. Bethlenfalvay. 1987. Mycorrhizal fungi and the integration of plant and soil nutrient dynamics. *J. Plant Nutr.* Vol. 10:1313-1321.

Ames, R.N., C.P.P. Reid, L.K. Porter, and C. Cambardella. 1983. Hyphal uptake and transport of nitrogen from two $_{15}$N-labelled sources by *Glomus mosseae*, a vesicular-arbuscular mycorrhizal fungus. *New Phytol.* 95:381-396.

Amijee, F. 1989. Colonization of roots by vesicular-arbuscular mycorrhizal fungi in relation to phosphorus and carbon nutrition. *Aspects Appl. Biol.* 22:219-226.

Azcón, R. 1989. Selective interaction between free-living rhizosphere bacteria and vesicular-arbuscular mycorrhizal fungi. *Biochem.* 21:639-644.

Azcón, R., R. Rubio, and J.M. Barea. 1991. Selective interactions between different species of mycorrhizal fungi and *Rhizobium meliloti* strains, and their effects on growth, N$_2$-fixation (^{15}N) and nutrition of *Medicago sativa* L. *New Phytol.* 117:399-404.

Azcón, R. and J.A. Ocampo. 1984. Effect of root exudation on VA mycorrhizal infection at early stages of plant growth. *Plant Soil* 82:133-138.

Azcón-Aguilar, C. and J.M. Barea. 1985. Effect of soil micro-organisms on formation of vesicular-arbuscular mycorrhizas. *Trans. Br. Mycol. Soc.* 84:536-537.

Azcón-Aguilar, C. and J.M. Barea. 1986. Effect of soil micro-organisms on spore germination and growth of the vesicular-arbuscular mycorrhizal fungus *Glomus mosseae*. *Trans. Br. Mycol. Soc.* 86:337-340.

Azcón-G. de Aguilar, C. and J.M. Barea. 1978. Effects of interactions between different culture fractions of 'phosphobacteria' and *Rhizobium* on mycorrhizal infection, growth, and nodulation of *Medicago sativa*. *Can. J. Microbiol.* 24:520-524.

Bagyaraj, D.J. 1984. Biological interactions with VA mycorrhizal fungi. p. 131-153. In: C.L. Powell and D.J. Bagyaraj (eds.), *VA mycorrhizas*. CRC Press, Inc. Boca Raton, FL.

Barea, J.M. 1991. Vesicular-arbuscular mycorrhizae as modifiers of soil fertility. *Adv. Soil Sci.* 15:1-40.

Barea, J.M., R. Azcón, and D.S. Hayman. 1975. Possible synergistic interactions between endogone and phosphate-solubilizing bacteria in low phosphate soils. p. 409-417. In: F.E. Sanders, B. Mosse, and P.B. Tinker (eds.), *Endomycorrhizas*. Academic Press, New York.

Barea, J.M., and C. Azcón-Aguilar. 1982. Production of plant growth-regulating substances by the vesicular-arbuscular mycorrhizae fungus *Glomus mosseae*. *Appl. Environ. Microbiol.* 43:810-813.

Barea, J.M. and C. Azcón-Aguilar. 1983. Mycorrhizas and their significance in nodulating nitrogen-fixing plants. *Adv. Agron.* 36:1-54.

Bécard, G., and J.A. Fortin. 1988. Early events of vesicular-arbuscular mycorrhiza formation on Ri T-DNA transformed roots. *New Phytol.* 108:211-218.

Bécard, G. and Y. Piché. 1989a. Fungal growth stimulation by CO_2 and root exudates in vesicular-arbuscular mycorrhizal symbiosis. *Appl. Environ. Microbiol.* 55:2320-2325.

Bécard, G., and Y. Piché. 1989b. New aspects on the acquisition of biotrophic status by a VAM fungus, *Gigaspora margarita. New Phytol.* 112:77-83.

Bethlenfalvay, G.J. and R.N. Ames. 1987. Comparison of two methods for quantifying extraradical mycelium of vesicular-arbuscular mycorrhizal fungi. *Soil Sci. Soc. Am J.* 51:834-837.

Bethlenfalvay, G.J. and R.L. Franson. 1989. Manganese toxicity alleviated by mycorrhizae in soybean. *J. Plant Nutr.* 12:953-970.

Bethlenfalvay, G.J., R.L. Franson, and M.S. Brown. 1990. Nutrition of mycorrhizal soybean evaluated by the Diagnosis and Recommendation Integrated System (DRIS). *Agron. J.* 82:302-304.

Bethlenfalvay, G.J., R.L. Franson, M.S. Brown, and K.L. Mihara. 1989. The *Glycine-Glomus-Bradyrhizobium* symbiosis. IX. Nutritional, morphological and physiological responses of nodulated soybean to geographic isolates of the mycorrhizal fungus *Glomus mosseae. Physiol. Plantarum* 76:226-232.

Bethlenfalvay, G.J., R.S. Pacovsky, H.G. Bayne, and A.E. Stafford. 1982. Interactions between nitrogen fixation, mycorrhizal colonization, and host-plant growth in the *Phaseolus-Rhizobium-Glomus* symbiosis. *Plant Physiol.* 70:446-450.

Bethlenfalvay, G.J., R.S. Thomas, S. Dakessian, M.S. Brown and R.N. Ames. 1988. Mycorrhizae in stressed environments: Effects on plant growth, endophyte development, soil stability and soil water. 1988. p. 1015-1029. In: E.E. Whitehead, C.F. Hutchinson, B.N. Timmermann, and R.G. Vandy (eds.), *Arid Lands Today and Tomorrow.* Westview Press, Boulder, CO.

Bevege, D.I. and G.D. Bowen. 1975. Endogone strain and host plant differences in development of vesicular-arbuscular mycorrhizas. p. 77-86. In: F.E. Sanders, B. Mosse, and P.B. Tinker (eds), *Endomycorrhizas.* Academic Press, N.Y.

Bolan, N.S. 1991. A critical review on the role of mycorrhizal fungi in the uptake of phosphorus by plants. *Plant Soil* 134:189-207.

Bowen, G.D. 1980. Mycorrhizal roles in tropical plants and ecosystem. p. 165-190. In: P. Mikoa (ed.), *Tropical mycorrhiza research.* Oxford Univ. Press, Oxford.

Bowen, G.D. 1987. The biology and physiology of infection and its development. p. 27-57. In: G.R. Safir (ed.), *Ecophysiology of VA mycorrhizal plants.* CRC Press, Inc., Boca Raton, FL.

Carr, G.R., M.A. Hinkley, F. LeTacon, C.M. Hepper, M.G.K. Jones, and E. Thomas. 1985. Improved hyphal growth of two species of vesicular-arbuscular mycorrhizal fungi in the presence of suspension-cultured plant cells. *New Phytol.* 101:417-426.

Coleman, D.C., C.P. Reid, and D.V. Cole. 1983. Biological strategies of nutrient cycling in soil systems. *Adv. Ecol. Res.* 13:1-55.

Cooper, K.M. and P.B. Tinker. 1981. Translocation and transfer of nutrients in vesicular-arbuscular mycorrhizas. IV. Effect of environmental variables on movement of phosphorus. *New Phytol.* 88:327-339.

Cox, G. and P.B. Tinker. 1976. Translocation and transfer of nutrients in vesicular-arbuscular mycorrhizas. I. The arbuscule and phosphorus transfer: A quantitative ultrastructural study. *New Phytol.* 77:371-378.

Duke, E.R., C.R. Johnson, and K.E. Koch. 1986. Accumulation of phosphorus, dry matter and betaine during NaCl stress of split-root citrus seedlings colonized with vesicular-arbuscular mycorrhizal fungi on zero, one or two halves. *New Phytol.* 104:583-590.

Evans, D.G. and M.H. Miller. 1988. Vesicular-arbuscular mycorrhizas and the soil-disturbance-induced reduction of nutrient absorption in maize. I. Causal relations. *New Phytol.* 110:67-74.

Evans, D.G. and M.H. Miller. 1990. The role of the external mycelial network in the effect of soil disturbance upon vesicular-arbuscular mycorrhizal colonization of maize. *New Phytol.* 114:65-71.

Fairchild, G.L. and M.H. Miller. 1988. Vesicular-arbuscular mycorrhizas and the soil-disturbance-induced reduction of nutrient absorption in maize. II. Development of the effect. *New Phytol.* 110:75-84.

Fairchild, G.L. and M.H. Miller. 1990. Vesicular-arbuscular mycorrhizas and the soil-disturbance-induced reduction of nutrient absorption in maize. III. Influence of P amendments to soil. *New Phytol.* 114:641-650.

Förster, I. and K. Freier. 1988. Contributions to the mobilization of phosphorus by soil microorganisms. 4th comm.: Investigation of the efficiency of P-mobilizing microorganisms in vitro and in the rhizosphere of sunflower and winter wheat. *Zentralbl. Mikrobiol.* 143:125-138.

Friese, C.F. and M.F. Allen. 1991. The spread of VA mycorrhizal fungal hyphae in the soil: Inoculum types and external hyphal architecture. *Mycologia* 83:409-418.

Furlan, V. and J.A. Fortin. 1973. Formation of endomycorrhize by *Endogone calospora* on *Allium cepa* under three temperature regimes. *Naturaliste Can.* 100:467-477.

Gemma, J.N. and R.E. Koske. 1988. Pre-infection interactions between roots and the mycorrhizal fungus *Gigaspora gigantia*: Chemotropism of germ tubes and root growth response. *Trans. Br. Mycol. Soc.* 91:123-132.

Gianinazzi, S. 1991. Vesicular-arbuscular (endo-) mycorrhizas: cellular, biochemical and genetic aspects. *Agric. Ecosystems Environ.* 35:105-119.

Gianinazzi, S., V. Gianinazzi-Pearson, B. Tisserant, and M.C. Lemoine. 1991. Protein activities as potential markers of functional endomycorrhizas in situ. *Third European Symposium on Mycorrhizas. Mycorrhizas in ecosystems — structure and function*, Sheffield, U.K.(Abstract).

Gildon, A. and P.B. Tinker. 1981. A heavy metal-tolerant strain of a mycorrhizal fungus. *Trans. Br. Mycol. Soc.* 77:648-649.

Graham, J.H. and J.P. Syvertsen. 1989. Vesicular-arbuscular mycorrhizas increase chloride concentration in citrus seedlings. *New Phytol.* 113:29-36.

Hamel, C., V. Furlan, and D.L. Smith. 1991. N_2-fixation and transfer in a field grown mycorrhizal corn and soybean intercrop. *Plant Soil* 133:177-185.

Harris, K.K. and R.E.J. Boerner. 1990. Effects of belowground grazing by collembola on growth, mycorrhizal infection, and P uptake of *Geranium robertianum*. *Plant Soil* 129:203-210.

Hendrix, P.F., R.W. Parmelee, D.A. Crossley, Jr., D.C. Coleman, E.P. Odum and P.W. Groffman. 1986. Detritus food webs in conventional and no-tillage agroecosystems. *BioScience* 36:374-380.

Hepper, C.M. 1984. Isolation and culture of VA mycorrhizal (VAM) fungi. p. 95-112. In: C.L. Powell and D.J. Bagyaraj (eds.), *VA mycorrhizae*. CRC Press, Inc. Boca Raton, FL.

Hepper, C.M. and I. Jakobsen. 1983. Hyphal growth from spores of the mycorrhizal fungus *Glomus caledonius*: effect of amino acids. *Soil Biol. Biochem.* 15:55-58.

Hepper, C.M. and A.Warner. 1983. Role of organic matter in growth of a vesicular-arbuscular mycorrhizal fungi in soil. *Trans. Br. Mycol. Soc.* 81:155-156.

Hetrick, B.A.D. 1984. Ecology of VA mycorrhizal fungi. p. 35-55. In: *VA mycorrhizae*. C.L. Powell and D.J. Bagyaraj (eds.). CRC Press, Inc. Boca Raton, FL.

Hetrick, B.A.D. 1986. Interactions of mycorrhizal fungi and other microbes. *Am. J. Bot.* 73:693.

Hetrick, B.A.D., G.W.T. Wilson, and J.F. Leslie. 1991. Root architecture of warm- and cool-season grasses: relationship to mycorrhizal dependence. *Can. J. Bot.* 69:112-118.

Hirrel, M.C. 1981. The effect of sodium and chloride salts on the germination of *Gigaspora margarita*. *Mycologia* 73:610-617.

Jakobsen, I. and L. Rosendahl. 1990. Carbon flow into soil and external hyphae from roots of mycorrhizal cucumber plants. *New Phytol.* 115:77-83.

Jasper, D.A., L.K. Abbott, and A.D. Robson. 1989. Soil disturbance reduces the infectivity of external hyphae of vesicular-arbuscular mycorrhizal fungi. *New Phytol.* 112:93-99.

Jasper, D.A., L.K. Abbott, and A.D. Robson. 1991. The effect of soil disturbance on vesicular-arbuscular mycorrhizal fungi in soils from different vegetation types. *New Phytol.* 118:471-476.

Jasper, D.A., A.D. Robson, and L.K. Abbott. 1979. Phosphorus and the formation of vesicular-arbuscular mycorrhiza. *Soil Biol. Biochem.* 11:501-505.

Johnson, N.C., F.L. Pfleger, R.K. Crookston, S.R. Simmons, and P.J. Copeland. 1991. Vesicular-arbuscular mycorrhizas respond to corn and soybean cropping history. *New Phytol.* 117:657-663.

Kaiser, P.A. and J. Lussenhop. 1991. Collembolan effects on establishment of vesicular-arbuscular mycorrhizae in soybean (*Glycine max*). *Soil Biol. Biochem.* 23:307-308.

Koske, R.E. 1982. Evidence for a volatile attractant from plant roots affecting germ tubes of VA mycorrhizal fungus. *Trans. Br. Mycol. Soc.* 79:305-310.

Kothari, S.K., H. Marschner, and V. Romheld. 1991. Effect of a vesicular-arbuscular mycorrhizal fungus and rhizosphere micro-organisms on manganese reduction in the rhizosphere and manganese concentrations in maize (*Zea mays* L.). *New Phytol.* 117:649-655.

Levy, Y., J.P. Syvertsen, and S. Nemec. 1983. Effect of drought stress and vesicular-arbuscular mycorrhiza on citrus transpiration and hydraulic conductivity of roots. *New Phytol.* 93:61-66.

Linderman, R.G. 1988. Mycorrhizal interactions with the rhizosphere microflora: The mycorrhizosphere effect. *Phytopathol.* 78:366-371.

Lynch, J.M. 1982. Microbe plant interaction in the rhizosphere. p. 395-411. In: R.G. Burns and J.H. Slater (eds.), *Experimental Microbial Ecology.* Blackwell Scientific Publishers, Oxford.

Macfadyen, A. 1969. The systematic study of soil ecosystems. p. 191-197. In: J.G. Sheals (ed.), *Ecosystem.* Systematics Assoc. Publ. No. 8. The Systematics Association, London.

Mayo, K., R.E. Davis, and J. Motta. 1986. Stimulation of germination of spores of *Glomus versiforme* by spore-associated bacteria. *Mycologia.* 78:426-431.

McGee, P.A. 1987. Alteration of growth of *Solanum opacum* and *Plantago drummondii* and inhibition of regrowth of hyphae of vesicular-arbuscular mycorrhizal fungi from dried root pieces by manganese. *Plant Soil* 101:227-233.

McGonigle, T.P., D.G. Evans, and M.H. Miller. 1990. Effect of degree of soil disturbance on mycorrhizal colonization and phosphorus absorption by maize in growth chamber and field experiments. *New Phytol.* 116:629-636.

Menge, J.A. 1982. Effect of soil fumigants and fungicides on vesicular-arbuscular fungi. *Phytopathol.* 72:1125-1132.

Meyer, J.R. and R.G. Linderman. 1986. Selective influence on populations of rhizosphere or rhizoplane bacteria and actinomycetes by mycorrhizas formed by *Glomus fasciculatum*. *Soil Biol. Biochem.* 18:191-196.

Miller, M.R. 1987. The ecology of vesicular-arbuscular mycorrhizae in grass- and shrublands. p. 135-170. In: G.R. Safir (ed.), *Ecophysiology of VA mycorrhizal plants*. CRC Press, Inc. Boca Raton, FL.

Moore, J.C., T.V. St. John, and D.C. Coleman. 1985. Ingestion of vesicular-arbuscular mycorrhizal hyphae and spores by soil microarthropods. *Ecol.* 66:1979-1981.

Morandi, D., J.A. Bailey, and V. Gianinazzi-Pearson. 1984. Isoflavonoid accumulation in soybean roots infected with vesicular-arbuscular mycorrhizal fungi. *Physiol. Plant Pathol.* 24:357-364.

Morandi, D. and V. Gianinazzi-Pearson. 1986. Influence of mycorrhizal infection and phosphate nutrition on secondary metabolic content of soybean roots, p. 787-791. In: V. Gianinazzi-Pearson and S. Gianinazzi (eds.), *Physiological and genetical aspects of mycorrhizae.* INRA Press, Paris.

Morgan, P., C.J. Cooper, and N.S. Battersby. 1991. Automated image analysis method to determine fungal biomass in soils and on solid matrices. *Soil Biol. Biochem.* 23:609-616.

Morton, J.B. 1990a. Evolutionary relationships among endomycorrhizal fungi of the Endogonaceae. *Mycologia* 82:192-207.

Morton, J.B. 1990b. Taxonomy of VA mycorrhizal fungi: classification, nomenclature, and identification. *Mycotaxon.* 32:267-324.

Morton, J.B. and G.L. Benney. 1990. Revised classification of arbuscular mycorrhizal fungi (Zygomycetes): A new order, Glomales, two new suborders, Glomineae and Gigasporineae, and two new families, Acaulosporaceae and Gigasporaceae, with emendation of Glomaceae. *Mycotaxon* 37:471-491.

Mugnier, J. and B. Mosse. 1987. Spore germination and viability of a vesicular-arbuscular mycorrhizal fungus, *Glomus mosseae. Trans. Br. Mycol. Soc.* 88:411-413.

Nair, M.G., G.R. Safir, and J.O. Siqueira. 1991. Isolation and identification of vesicular-arbuscular mycorrhizal-stimulatory compounds from clover (*Trifolium repens*) roots. *Appl. Environ. Microbiol.* 57:434-439.

Nelsen, C.E. 1987. The water relations of vesicular-arbuscular mycorrhizal systems. p. 71-91. In: G.R. Safir (ed.), *Ecophysiology of VA mycorrhizal plants.* CRC Press, Boca Raton, FL.

Newman, E.I., and H.J. Bowen. 1974. Patterns of distribution of bacteria on root surfaces. *Soil Biol. Biochem.* 6:205-209.

Ocampo, J.A. and D.S. Hayman. 1981. Influence of plant interactions on vesicular-arbuscular mycorrhizal infections. II. Crop rotations and residual effects of non-host plants. *New Phytol.* 87:333-343.

Odum, E.P. 1984. Properties of agroecosystems. p. 5-11. In: R. Lowrance, B.R. Stinner, and G.J. House (eds.), *Agricultural ecosystems.* J. Wiley & Sons, N.Y.

Oliver, A.J., S.E. Smith, D.J.D. Nicholas, W. Wallace, and F.A. Smith. 1983. Activity of nitrate reductase in Trifolium Subterraneum: Effects of mycorrhizal infection and phosphate nutrition. *New Phytol.* 94:63-79.

O'Neill, E.G., R.V. O'Neill, and R.J. Norby. 1991. Hierarchy theory as a guide to mycorrhizal research on large-scale problems. *Environ. Pollution* 73:271-284.

Patten, B. C. and E. P. Odum. 1981. The cybernetic nature of ecosystems. *Amer. Naturalist* 118:886-895.

Paula, M. D. and J. O. Siqueira. 1990. Stimulation of hyphal growth of the VA mycorrhizal fungus Gigaspora margarita by suspension-cultured *Pueraria phaseoloides* cells and cell products. *New Phytol.* 115:69-75.

Rabatin, S.C. and B.R. Stinner. 1985. Arthropods as consumers of vesicular-arbuscular mycorrhizal fungi. *Mycologia* 77:320-322.

Safir, G.R., J.S. Boyer, and J.W. Gerdemann. 1972. Nutrient status and mycorrhizal enhancement of water transport in soybean. *Plant Physiol.* 49:700-703.

Sanders, F.E. 1975. The effect of foliar-applied phosphate on the mycorrhizal infections of onion roots. p. 261-276. In: F.E. Sanders, B. Mosse, and P.B. Tinker (eds), *Endomycorrhizas*. Academic Press, New York.

Sanders, F.E. and B.P. Tinker. 1973. Phosphate flow into mycorrhizal roots. *Pestic. Sci.* 4:385-395.

Sattar, M.A. and A.C. Gaur. 1987. Production of auxins and gibberellins by phosphate-dissolving microorganisms. *Zentralbl. Microbiol.* 142:393-395.

Schenck, N.C., S.O. Graham, and N.E. Green. 1975. Temperature and light effect on contamination and spore germination of vesicular-arbuscular mycorrhizal fungi. *Mycologia* 67:1189-1192.

Schubert, A., C. Marzachi, M. Mazzitelli, M.C. Cravero, and P. Bonfante-Fasolo. 1987. Development of total and viable extraradical mycelium in the vesicular-arbuscular mycorrhizal fungus *Glomus clarum* Nicol. & Schenck. *New Phytol.* 107:183-190.

Schwab, S.M., J.A. Menge, and R.T. Leonard. 1983. Comparison of stages of vesicular-arbuscular mycorrhiza formation in Sudan grass grown at two levels of phosphorus nutrition. *Amer. J. Bot.* 70:1225-1232.

Schwab, S.M., J.A. Menge, and P.B. Tinker. 1991. Regulation of nutrient transfer between a fungus and vesicular-arbuscular mycorrhiza. *New Phytol.* 117:387-398.

Schwab, S.M., R.T. Leonard, and J.A. Menge. 1984. Quantitative and qualitative comparison of root exudates of mycorrhizal and nonmycorrhizal plant species. *Can. J. Bot.* 62:1227-1231.

Siqueira, J.O., D.M. Sylvia, J. Gibson, and D.H. Hubbell. 1985. Spores, germination, and germ tubes of vesicular arbuscular mycorrhizal fungi. *Can. J.Microbiol.* 31:965-971.

Siqueira, J.O., G.R. Safir, and M.G. Nair. 1991. Stimulation of vesicular-arbuscular mycorrhiza formation and growth of white clover by flavonoid compounds. *New Phytol.* 118:87-93.

Smith, S.E. and G.D. Bowen. 1979. Soil temperature, mycorrhizal infection and nodulation of *Medicago trunculata* and *Trifolium subterraneum*. *Soil Biol. Biochem.* 11:469-473.

Smith, S.E. and V. Gianinazzi-Pearson. 1988. Physiological interactions between symbionts in vesicular-arbuscular mycorrhizal plants. *Ann. Rev. Plant Physiol. Plant Mol. Biol.* 39:221-244.

Smith, S.E. and N.A. Walker. 1981. A quantitative study of mycorrhizal infection in trifolium: separate determination of the rates of infection and of mycelial growth. *New Phytol.* 89:225-240.

Smith, S.E., N.A. Walker, and M. Tester. 1986. The apparent width of the rhizosphere of *Trifolium subterraneum* L. for vesicular-arbuscular mycorrhizal infection: Effects of time and other factors. *New Phytol.* 104:547-558.

St. John, T.V., D.C. Coleman, and C.P.P. Reid. 1983. Association of vesicular-arbuscular mycorrhizal hyphae with soil organic particles. *Ecol.* 64:957-959.

Strullu, D.-G., C. Romand, and C. Plenchette. 1991. Axenic culture and encapsulation of the intraradical forms of Glomus spp. *World J. Microbiol. Biotechnol.* 7:292-297.

Sweatt, M.R. and F.T. Davies, Jr. 1984. Mycorrhizae, water relations, growth, and nutrient uptake of geranium grown under moderately high phosphorus regimes. *J. Amer. Soc. Hort. Sci.* 109:210-213.

Sylvia, D.M. 1988. Activity of external hyphae of vesicular-arbuscular mycorrhizal fungi. *Soil. Biol. Biochem.* 20:39-43.

Sylvia, D.M. 1990. Distribution, structure, and function of external hyphae of vesicular-arbuscular mycorrhizal fungi, p. 144-167. In: J.E. Box, Jr. and L.C. Hammond (eds.), *Rhizosphere dynamics*. AAAS Selected Symposia Series, Westview Press, Inc. CO.

Sylvia, D.M. and L.H. Neal. 1990. Nitrogen affects the phosphorus response of VA mycorrhiza. *New Phytol.* 115:303-310.

Thompson, J.P. 1986. Soilless culture of vesicular-arbuscular mycorrhizae of cereals: effects of nutrient concentration and nitrogen source. *Can. J. Bot.* 64:2282-2294.

Thompson, J.P. 1987. Decline of vesicular-arbuscular mycorrhizae in long fallow disorder of field crops and its expression in phosphorus deficiency of sunflower. *Aust. J. Agric. Res.* 38:847-867.

Tylka, G.L., R.S. Hussey, and R.W. Roncadori. 1991. Interactions of vesicular-arbuscular mycorrhizal fungi, phosphorus, and *Heterodera glycines* on soybeans. *J. Nematol.* 23:122-133.

Van Kessel, C., P.W. Singleton, and H.J. Hoben. 1985. Enhanced N-transfer from a soybean to maize by vesicular-arbuscular mycorrhizal (VAM) fungi. *Plant Physiol.* 79:562-563.

Warner, A. 1984. Colonization of organic matter by vesicular-arbuscular mycorrhizal fungi. *Trans. Br. Mycol. Soc.* 82:352-354.

Wilson, D.O. 1988. Differential plant response to inoculation with two VA mycorrhizal fungi isolated from a low-pH soil. *Plant Soil* 110:69-76.

Earthworms and other Fauna in the Soil

Edwin C. Berry

I. Introduction

Soil organisms range in size from protozoa which may be less than 5 μm in length and weigh $< 10^{-10}$ gm to earthworms, which may attain lengths of > 1 meter and diameters of > 20 mm and weights in excess of 500 gms (Lee and Foster, 1991). These organisms are found within the soil matrix occupying existing pore spaces as well as within the litter layer. Soil arthropods have been arranged and separated into groups based upon their associations with the soil. Eisenbeis and Wichard (1985) distinguished three groups which consisted of euedaphon, epedaphon, and hemiedaphon. Euedaphic forms primarily inhabit the lower soil layers in existing pore systems. Their diameter is restricted to the size of the pores in this region. Organisms in these regions are usually round or have worm-like body forms. Epedaphic forms live on the soil surface and in the litter layer. Epedaphic organisms are not restricted in size to conform to pore diameters, and body forms are highly variable. Hemiedaphic forms do not

This chapter was prepared by a U.S. government employee as part
of his official duties and legally cannot be copyrighted.

occupy positions between the euedaphic and epedaphic forms. Usually the hemiedaphic form is a temporary form of life enabling the organisms to occupy existing, or more often, self-made burrows in the soil during unfavorable environmental conditions, for example, to avoid desiccation and cold.

Swift et al. (1979) recognized that body width, from a functional standpoint, was more important in defining the extent in which soil animals are constrained by, or modify, the structure of soil and litter habitats. Accordingly, soil invertebrates were divided into three classes: micro- ($< 100 \ \mu m$), meso- (< 1-2 mm) and macro-fauna (< 10-20 mm). The microfauna (primarily Protozoa and Nematoda) are primarily associated with waterfilms and water filled voids and do not modify the structure of soils nor contribute to the formation of particulate organic matter (Anderson, 1988). Mesofauna consist of enchytraeids, Collembola (insects), and Acarina (mites), and are primarily associated with air filled pores. This group feeds primarily on organic matter and fungal hyphae associated with decomposing organic matter. Fecal matter from this group contribute aggregates to the surface horizons. The macrofauna include isopods, millipedes, earthworms, ants, and termites and are considered to be more important than the micro- and meso-fauna in affecting soil structure through their feeding and burrowing habits.

Further distinction between the groups exists in the movement of mineral soils within the excreted material. The macro-arthropods ingest or utilize large amounts of the mineral soil in feeding and nest building whereas microarthropods do not include any of the soil mineral materials in food gathering and nest building.

Hole (1981) summarized the biological activities of the soil fauna as mounding, mixing, forming macropores, forming and destroying peds, and regulating nutrient cycling. These activities facilitate movement of air and water in the soil matrix and control soil erosion and decomposition of plant and animal litter. Through the activities of the soil fauna the physical, chemical, and biological characteristics of the soil and its ecosystem are generated and maintained.

Soil structure may be modified directly or indirectly by the soil fauna. Direct effects result from burrowing, redistribution of residue, and deposition of feces containing undigested residue and mineral soil. Activities of the soil fauna influence transport of materials indirectly by changing water movement on and in the soil, formation of particulate, and soluble materials that are transported by water, wind, and gravity (Anderson, 1988).

Air and water movement in the soil are also influenced by the complex burrowing and feeding habits of the soil fauna. Burrowing is primarily associated with searching for food, activities associated with reproductive behaviors, and escaping unfavorable environmental conditions. Orientation, size, and opening to the surface of the burrows are important in water and air movement through the soil matrix. Burrow orientation may be either vertical or horizontal and can be open to the surface. Burrows can be closed by egested materials. Burrow formation and function will be discussed in greater detail in Section II.

Soil animals participate as consumers and aid in the cycling of nutrients and energy. Their harvesting and food storage activities result in the transference of nutrients to their nests and burrows. Soil animals can be classified as primary or secondary consumers. Primary consumers, such as insect larvae, feed on roots of primary producers (plants). Most soil animals are secondary consumers and feed on dead and decaying organic material. Soil organisms in this category are represented by Collembola (springtails), mites, slugs, earthworms, and various soil insects.

It is accepted that energy and nutrients obtained by plants are usually tied up in plant residue at the end of the growing season (Reiners, 1973). Mineralization must occur before this material can be returned to the ecosystem. Soil fungi and bacteria are directly responsible for most of the decomposition.

Protozoan, nematodes, earthworms, and arthropods influence the functioning of the decomposer flora as a direct or indirect result of their feeding (Seastedt and Crossley, 1984). Soil fauna are responsible for controlling nutrient cycling (Coleman et al., 1984), regulation of micro- and meso-fauna in the soil matrix (Visser, 1985), soil genesis (Seastedt, 1984), and energy dynamics of the soil (Peterson and Luxton, 1982).

Decomposition and nutrient cycling are determined by resource quality and accessibility, temperature and moisture, and the available decomposer community (Parkinson, 1988). Mites and Collembola generally account for about 95% of the total microarthropods (Harding and Stuttard, 1974). Densities of these micro-arthropods range from 50,000 m^{-2} in the tropics (Madge, 1969) under low levels of organic matter to 300,000 m^{-2} in sites where extensive amounts of aerobic organic matter occur (Persson et al., 1980).

The relationship between microbes and other fauna communities is complex and the processes and mechanisms in organic matter decomposition and nutrient cycling are poorly understood. Visser (1985) summarized the mechanisms by which soil fauna influence the composition of microbial communities as follows: 1) comminution, channelling and mixing may reduce fungal species numbers and divert fungal successional patterns on decaying plant residues; 2) grazing by fauna on selected fungi may alter fungal distribution and succession on decomposing plant debris; and 3) dispersal of soil micro-organisms. Fragmentation of organic matter, channelling, and mixing of soil components are key roles by which the soil fauna stimulate microbial activity and enhance the rate of organic matter decomposition (Crossley, 1977a; Swift et al., 1979). In general, the non-ingested fragments are larger than the ingested fragments of organic matter and the degree of reduction depends on the size of the invertebrate. According to Swift et al. (1979), plant debris is reduced to particles 200-300 μm in diameter by earthworms and millipedes, 100-200 μm by medium-sized arthropods and 20-50 μm by micro-arthropods.

The objectives of this chapter are to summarize the effects of earthworms and other soil fauna on soil properties and residue decomposition. Due to the broad range of soil biota, this review will be limited primarily to earthworms, Collembola, mites, ants, and termites.

II. Earthworms

The importance of earthworms was first recognized by Darwin (1881) who stated, "The plough is one of the most ancient and most valuable of man's inventions; but long before he existed the land was in fact regularly ploughed and still continues to be thus ploughed by earthworms. It may be doubted whether there are many other animals which have played so important a part in the history of the world as have these lowly organized creatures." Since Charles Darwin's time, research on the activities associated with earthworms has increased.

Benefits from the activities of earthworms have been summarized by many authors and include:

1) burrowing and excavation while searching for food, providing living and estivation sites, and cocoon deposition,
2) lining of burrows with mucus, fragmented residue, and excreta,
3) ingestion of soil materials including mineral soil and plant residue,
4) transporting plant residue from the soil surface to within the soil matrix,
5) and transporting mineral soil material or ingested plant material to the soil surface or within voids within the soil matrix.

These activities directly or indirectly affect soil genesis, soil texture as well as consistency, soil aggregate stability, macroporosity, water infiltration, gas exchange, and residue decomposition.

A. Classification and Ecological Significance

Unfortunately, much of the literature assumes that all species of earthworms have similar effects on modifying soil structure and enhancing residue decomposition. Because food gathering and burrowing activities vary among species, it should not be assumed that all species of earthworms affect soils in identical manners. Several classification systems have been proposed to place similar species within ecological groups. These systems are based on vertical distribution within the soil, color, reproductive activities, food source, and habitats.

Evans (1948) as well as Evans and Guild (1947a, 1948) divided the British Lumbricidae into two distinct groups based on habitat, burrow formation, casting activity, and the ability to survive adverse environmental conditions by entering a dormant state (estivation). According to their classification, one group consisted of lumbricids that are surface-dwellers and do not live in burrows. Species within this group do not produce recognizable casts and do not aestivate. The other group consisted of earthworms that maintain burrows deep within the soil matrix. Casts are produced on the surface of the soil or in voids within the soil. Unfavorable environmental conditions are avoided by estivation.

Graff (1953) expanded the classification system proposed by Evans and included coloration, cocoon production, rate of maturity, and fecundity. According to this system, surface-dwelling earthworms were usually red in color and predominantly occurred in habitats with organic surface horizons. Individuals within this group matured rapidly, produced many cocoons, and had several generations per year. Earthworms in the soil-dwelling group were mostly lacking in pigmentation. Species within the soil dwelling group matured slowly, had low rates of cocoon production, and usually had only one generation per year.

Lee (1959) described three functional groups among New Zealand megascolecoids. His grouping was primarily based on relation of the earthworms to the soil horizons in which they were found. Lee's groups consisted of those earthworms that were found most often in the leaf mold (litter or O horizon), species in the top soil (A horizon), or those found in the subsoil (B and/or B/C horizon). Lee further defined the groups on the basis of morphological similarities, behavioral and physiological features, food preference, susceptibility to predation, geographical distributions, and reaction to changes in land use patterns.

Later, Bouché (1977) recognized three major morpho-ecological groups among the European Lumbricidae. Bouché distinguished the groups primarily on the basis of morphological characters that were important in relation to function. Bouché's groups consisted of *epigeic*, *anecic*, and *endogeic* species. Epigeic species live and feed in the litter layer and do not make permanent burrows. The anecic species live in the mineral soil but feed on leaf litter on the soil surface. Species within this group maintain relatively permanent vertically oriented burrows. The endogeic species live and feed below the soil surface. Endogeic species have extensive systems of burrows that are more or less oriented horizontally and rarely open to the surface. Endogées are typically represented by *Aporrectodea trapezoides* (Duges), *A. tuberculata* (Eisen), *A. turgida* (Eisen), and *A. rosea* (Savigny) whereas the anéciques are represented by *Lumbricus terrestris* (Linneaus) and *L. rubellus* (Hoffmeister). Among the three groups, earthworms in the endogées and anéciques are probably more common and important in agriculture row crops (Kladivko and Timmenga, 1990).

Lumbricids have also been separated and classified on the basis of the state of decomposition of their food source. Perel (1977) distinguished two groups which consisted of *humus formers* and *humus feeders*. This system of classification is particularly useful in examining the effects of earthworms on residue decomposition and nutrient cycling. Earthworms in the humus forming group feed on plant debris that is slightly decomposed and are typically represented by *L. terrestris* and *L. rubellus*. Both species feed on relatively non-decomposed plant material on the soil surface. In contrast, earthworms in the humus feeding

Table 1. Earthworm populations in selected habitats from various locations

Habitat	Number m⁻²	Reference
Alfalfa	258-1270	Slater and Hopp, 1947
Soybeans	140-667	Slater and Hopp, 1947
Corn	5-150	De St. Remy and Daynard, 1982
Small grain	84	Edwards and Lofty, 1982
Pasture	400-500	Cotton and Curry, 1980
Natural grassland	13-41	Reynolds, 1970
Orchards	300-500	Rhee, van and Nathans, 1961
Corn	22-65	Tomlin and Miller, 1988
Forest	14-124	Reynolds, 1972
Forest	240-780	Maldague, 1970

group (represented by *A. caliginosa)* feed on more highly decomposed plant material that is dispersed in the soil mineral layer (Piearce, 1978).

Some species may occur in more than one category depending upon stage of development, food availability, and soil conditions. For example, *L. badensis* is usually considered an anecic species during the adult stage, but may live epigeically during the juvenile stage (Kobel-Lamparski and Lamparski, 1987). Cocoon chambers are formed by adult *L. badensis* in the soil matrix at depths between 0.4 and 1.5 m. Juvenile worms migrate to the soil surface through the adult worm burrows and live in horizontal tubes during the first growing season (epigeic classification). After the first season, the juvenile *L. badensis* form several U-shaped burrows. Later a main tube is made which penetrates into the soil and may extend to 2.5 m (anecic classification).

L. terrestris is usually considered an anécique, but may live as an épigee. In forest habitats, *L. terrestris* live in the litter layer and do not make permanent burrows (epigeic) and in other habitats, they maintain permanent, vertical burrows (Bouché, 1972 [quoted in Kobel-Lamparski and Lamparski, 1987]). Similarly, *L. rubellus* may live either way. In habitats with a litter layer, this species lives as an épigee whereas in open areas, juveniles and adults convert to an anecic way of life (Kobel-Lamparski and Lamparski, 1983).

B. Distribution and Abundance

Climatic conditions, regional variation in vegetation, food sources, and soil texture are primarily responsible for determining the abundance and species diversity of earthworm communities. Data presented in Table 1 demonstrates that earthworms are usually more abundant in habitats that are not disturbed. However, interpretation of data from various locations is difficult since not all population estimates were obtained with the same sampling procedure.

Lee (1985) and Satchell (1983) summarized data on the abundance and biomass of earthworms from several habitats throughout the world. Their data show that earthworms are lacking only in those regions of extreme cold or drought and in habitats with high salinity or ones deficient in organic matter.

C. Population Determinants

Earthworm populations may be influenced by the physical and chemical characteristics of the soil, type of vegetation, and land management practices. However, in most cases the primary determinants are the supplies of available moisture and food. In many situations, earthworms populations may be prevented by unfavorable moisture and temperature conditions even if a suitable food source is available (Satchell, 1967). Earthworm growth and development may also be influenced not only by the quantity but also by the quality of crop residue (Lofs-Holmin, 1983). Transitions from permanent grassland to cultivated land are accompanied by declines in earthworm numbers (Barnes and Ellis, 1979). This decline is usually contributed to the loss of soil organic matter. However, it is not known what qualities and quantities of organic matter are needed to support populations of different species of earthworms in different crops.

In Sweden, Bostrom and Lof-Holmin (1986) studied the growth of juvenile *Allolobophora caliginosa* when fed shoots and roots of barley, meadow fescue, and alfalfa. Their studies included growth of the juveniles in relation to particle size, protein, crude fiber content, and toxicities of fresh residue. Juvenile *A. caliginosa* showed extremely different growth rates when fed different plant materials. Growth rates were greater on fescue shoots than lucerne shoots or roots or fescue roots. Growth on barley was slow and the final mass of *A. caliginosa* was less than when grown on the other plant materials. In addition, Bostrom and Lof-Holmin reported that particle size was an important factor in governing growth whereas protein and crude fiber contents of the crops were not correlated to growth rate. They suggested that the smaller plant particles probably enabled digestion, resulting in an increased consumption of microorganisms which provide additional nutrient sources. They concluded that fine roots and the small-particle fraction of residues remaining after harvest are more important food sources than straw for earthworms. The importance of particle size was also demonstrated by Neuhauser et al. (1980) with *Eisenia foetida*. In their studies, growth rates were greater when *E. foetida* fed on particles 0.3 mm^2 than on particle sizes of 1.0 mm^2.

Food availability is a prime factor in regulating earthworm biomass and abundance (Satchell, 1967). However, moisture and temperature may limit the ability of the earthworms to exploit these resources. Under adverse moisture or temperatures, earthworms respond by either entering a state of estivation or migrating to lower depths in the soil. In either case, feeding and reproduction ceases until favorable environmental conditions prevail.

Other researchers have suggested that as earthworms feed they ingest bacteria and fungi which are necessary for growth (van Rhee, 1963). In order to investigate the importance of the state of decomposition of the plant material, Bostrom (1987) compared barley, lucerne and meadow fescue in various states of decomposition (as determined by C/N ratios). Plant material was buried in jute cloth bags for various intervals of time before being fed to juvenile *Allolobophora caliginosa*. Juveniles grew more rapidly on meadow fescue and the slowest on barley, and growth rates were lower in fresh material than in material that was buried and allowed to decompose for three months. Earthworms remained active, however, for a longer period of time in barley than in the other materials. Bostrom suggested that since barley has a 20% lignin content as compared to less than 10% for fescue and lucerne, decomposition was slower in the barley and this prolonged the earthworm activity. In this study, fresh mixtures of lucerne were toxic to the juveniles. However, toxicity could not be explained by concentrations of nitrates, nitrites, or ammonium. It was concluded by Bostrom that some substance present in the lucerne or a substance formed by the microorganisms associated with the lucerne may have retarded the growth in the non-decomposed lucerne.

Earthworm populations are affected by agricultural management systems. The reader should refer to more comprehensive reviews by Edwards and Lofty (1977), Lee (1985), and Kladivko and Timmenga (1990) on the effects of tillage and cropping on earthworm communities. Generally, higher earthworm populations are favored by cropping systems in which the returning crop residue provides food and a continuous ground cover. Tillage alters the soil-litter conditions. As tillage decreases, the amount of residue from the previous crop increases which creates microenvironmental conditions favorable for soil fauna (House and Stinner, 1983). Comparisons between no-tillage and conventional tillage have shown that when tillage decreases, earthworm populations generally increase (Barnes and Ellis, 1979; Edwards and Lofty, 1982; Clutterbuck and Hodgson, 1984; Mackay and Kladivko, 1985; De St. Remy and Daynard, 1982; House and Parmelee, 1985).

Limited population data are available for earthworms in agroecosystems in North America. Data presented in Table 2 show the effects of habitats, plant species, and cropping systems on earthworm populations in North America.

Populations varied from 8 earthworms m^{-2} in Indiana to 2,202 earthworms m^{-2} in Georgia. Low population estimates (Mackay and Kladivko, 1985) were recorded from a long term tillage experiment (8 years) comparing the effect of no-till and conventional tillage (mold board plow) practices in continuous soybean and corn production. Sample sites were in a Chalmers silt loam (Typic Haplaquoll). Mackay and Kladivko (1985) speculated that the low numbers of earthworms in the plowed corn fields were influenced by crop type, use of insecticides (terbufos), anhydrous ammonia fertilizer, or the herbicides (atrazine and cyanazine).

The greatest abundance of earthworms were reported by House and Parmelee (1985) from an experiment comparing soil arthropods and earthworms from

Table 2. Earthworm populations in selected habitats in North America

Habitat	Tillage	Sample method	Number m^{-2}	Source
Corn	No till	Hand sorting	190	Berry and Karlen, 1983
Corn	No till	Hand sorting	16	Mackay and Kladivko, 1985
Corn	No till		48	De St. Remy and Daynard, 1982
Corn	No till	Hand sorting	34-55	Fuchs and Linden, 1988
Alfalfa	No till	Electric probe	367	Kemper et. al., 1987
Corn	Tilled		55	De St. Remy and Daynard, 1982
Corn	Tilled	Hand sorting	8	Mackay and Kladivko, 1985
Corn	Tilled	Hand sorting	73	Berry and Karlen, 1993
Corn	Tilled	Hand sorting	51-58	Fuchs and Linden, 1988
Soybean	No till	Hand sorting	141	Mackay and Kladivko, 1985
Soybean	Tilled	Hand sorting	62	Mackay and Kladivko, 1985
Soybean	No till	Hand sorting	267	Fuchs and Linden, 1988
Pasture		Hand sorting	1298	Mackay and Kladivko, 1985
Grasses	No till	Hand sorting	363	Fuchs and Linden, 1988
Sorg/rye	No till	Hand sorting	2202	House and Parmelee, 1985
Sor/clov	No till	Hand sorting	1210	House and Parmelee, 1985
Alf/gras			205	De St. Remy and Daynard, 1982
Alfalfa		Hand sorting	827	Fuchs and Linden, 1988

conventional and no-tillage agroecosystems, Hiwassee loam (Typic Rodudult), in Georgia. Although the authors did not speculate on the reasons for the large numbers of earthworms in the no till plots, it was probably due to the lack of soil disturbance by plowing, disking, and cultivation for 17 years. The lack of insecticide usage may have also contributed to the extremely large earthworm populations.

Negative effects of tillage on earthworm populations have been contributed to mechanical damage to earthworms during cultivation, loss of ground cover that would buffer climatic changes, and decreases in food supply (Edwards and Lofty, 1982). Data on the effects of tillage on earthworm populations are indirect and may be more indicative of system changes rather than direct mechanical effects. Because earthworms may recover from severe wounds by regeneration, it is not likely that mechanical damage accounts for population reductions.

After Ontario growers voiced concern over possible earthworm eradication in plowed fields, Tomlin and Miller (1988) investigated the impact of ring-billed gull, *Larus delawarensis* Ord., foraging activity on earthworm populations in Southwestern Ontario agricultural soil. Estimates of the numbers of exposed earthworms were determined by collecting individuals within one minute after plowing at five sites. These data later showed that an average of 11% (12 earthworms m^{-2}) of the initial population (111 earthworms m^{-2}) were exposed. However, 37% (4.5 earthworms m^{-2}) of the population that were exposed by plowing were injured by the mechanical action of the plow. Information regarding how many worms would have died from the plowing action or information regarding regeneration was not included in this study.

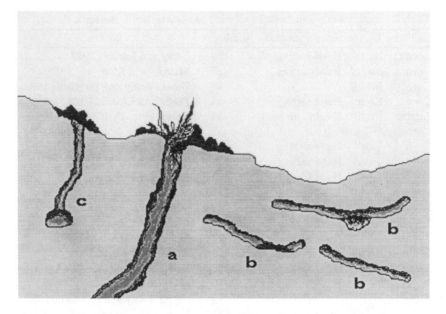

Figure 1. Burrows of different species of earthworms in the soil. (a) typical burrow shapes of surface-feeding species; (b) burrows of subsurface-feeding species; and c) burrow of surface-dwelling species.

D. Burrows

Burrow diameters generally range from 1-10 mm and may have volumes of 1.3-9.6 liters per square meter (Kretzschmar, 1978, 1982). Densities of earthworm burrows have been estimated as high as 2,000 burrows m^{-2} by many researchers. Numbers of burrows vary with population densities, cropping practices, and geographic locations. For example, Bouché (1971) reported more than 800 burrows m^{-2} in a pasture soil. Tisdall (1978), in France, counted 2,000 burrows m^{-2} in an irrigated orchard soil.

Earthworms move in the soil matrix to search for food, favorable environmental conditions (moisture and temperature), and to colonize new environments. Burrow construction, orientation, opening to the surface, and time of occupation is species specific. Lee (1985) describes three principal types of burrows. The first type (Figure 1a) is usually occupied by anéciques species (for example, *L. terrestris*). This type of burrow is more or less permanent, extending down in the underlying soil horizon. These burrows are usually vertical but may branch at the surface forming several entrances. This type of burrow may extend several meters down into the soil. The walls of the burrows usually have smooth linings built up by compression of the soil and mucus secreted by the earthworm. The second type of burrow is temporary and is developed by those species that feed in the subsurface soil horizons (Figure 1b). These burrows are usually horizontal, but may have vertical components, and may or may not open to the surface. These burrows may be partly or wholly packed with casts or soil carried down from the upper horizons by water. The third type of burrow (Figure 1c) is more or less vertical, formed by earthworms that live near the soil surface. This type of burrow is usually used as a refuge when unfavorable conditions exist at the soil surface. These burrows lack distinct lining and may terminate in spherical mucus-lined chambers. Since they are temporary, they are often filled with casts and soil particles from the upper horizons.

Burrowing activity is species specific and is influenced by food source and size of individuals. Evans and Guild (1947b) observed that *L. terrestris* and *L. rubellus* constructed few burrows in the soil when ample dung was on the soil surface but when dung was scarce, extensive burrows were constructed. Using glass observation cages, Martin (1982) investigated the interaction between organic matter in soil and the burrowing activity of *A. trapezoides, A. liginosa,* and *L. rubellus.* Burrows produced by *A. caliginosa* were longer than those produced by *A. trapezoides*. They concluded that these differences were due to the size of the test specimen rather than differences between the two species. They cited similar results presented by Bolton and Phillipson (1976) for *A. rosea,* whereas smaller specimens of *A. rosea* formed longer burrows than did larger specimens. Martin's results with *L. rubellus* indicated that the source of food was important in burrow formation. In test containers with grass meal, worms formed only a few long burrows with many feeding caverns. In containers without food, several very long burrows with many short sided burrows were formed. Martin concluded that top-soil dwelling earthworms respond to decreasing food supplies by increasing consumption.

Burrowing activity may also be influenced by soil moisture and temperatures. Scheu (1987) investigated the influence of three temperatures (5, 10, and 15°C) and three soil moistures (48, 60, and 73% water of dry soil weight) on the length of burrows produced by *A. caliginosa* and *O. lacteaum.* The burrowing activity of *A. caliginosa* and *O. lacteaum* was affected by both temperature and soil moisture. In a soil of 60% water content, the burrows at 10°C were 1.6 times as long as the burrows constructed at 5°C and at 15°C they were 2.2 times as long.

Mechanisms for forming burrows are described by Lee (1985) and Lee and Foster (1991). Earthworms may form burrows in two ways. The anterior region of the body is inserted into spaces between soil particles and then the body is expanded to enlarge the burrow diameter. If the soil is compacted and neither cracks or crevices exist, burrows are formed by ingesting the soil. In the latter method, the pharynx is everted, soil is moistened by saliva, and the material is drawn into the gut with suction. The ingested soil is then excreted as casts on the soil surface or in voids within the burrows.

Dexter (1978) examined whether *Allolobophora caliginosa* pushed the soil aside as they tunneled forward or whether they ingest the soil ahead of them. Soil clods containing burrows produced by *A. caliginosa* were compared to clods in which artificial burrows were produced with a tapered plastic needle similar in size to the worm burrows. Radiographs from epoxy resin clods showed that with artificial burrows the soil was compacted around the burrow and lateral cracks occurred beyond the tunnel wall. However, burrows produced by *A. caliginosa* were without disturbance to the surrounding soil and lateral cracks were nonexistent. Dexter concluded that with this species the principal mechanism of tunneling is by ingestion of the soil.

Regardless of how burrows are formed, axial pressures for forward movement and radial pressures for removing the soil laterally must be exerted. McKenzie and Dexter (1988a, 1988b) measured these pressures exerted by *Aporrectodea rosea* as they moved through 10 mm thick discs. It was shown that *A. rosea* could exert axial pressures of approximately 100 kPa and radial pressures in excess of 230 kPa. McKenzie and Dexter summarized that these pressures are insufficient to allow penetration of most soil aggregates. Thus, construction of burrows is facilitated by penetration through entering existing soil voids or by ingestion of the soil.

Earthworm burrows are lined with a protein-rich mucus that serves to lubricate passage of the earthworm through the soil. Burrows are also lined with alleviated material and partially decomposed organic matter. Bouché (1975) recognized the 2 mm soil layer around the perimeter of the burrow as the drilosphere. Lee and Foster (1991) summarized the work of Jeanson (1961, 1964, 1971) on the nature of the burrow lining. Jeanson (1964) examined thin sections of the drilosphere from burrows of *L. terrestris* and *Aporrectodea icterica* in a loam from the textural B horizon of a loess soil. Burrows were lined with a thin continuous layer of oriented clay particles and outside this layer there were three concentric zones of humic material, calcium carbonate and iron oxides. Jeanson also showed that burrows formed by *L. terrestris* had even distributions of iron and manganese on the walls. Calcium, however, was unevenly distributed in "corpuscles" of calcite.

The nature of the burrow wall may vary with the food source of the earthworm. Jeanson (1961) examined burrows of *L. terrestris* that were reared in the laboratory on ground lucerne. Under these conditions, a blackish layer about 1 mm thick formed on the burrow walls. This layer consisted of soil mineral particles and humic material that was deposited on the burrow walls. Outside this layer was a layer of reddish ferric-iron, 2-3 mm thick, apparently formed by the oxidation of ferrous-iron compounds. Similar results were reported by Kretzschmar (1982) for *Aporrectodea longa* and *A. caliginosa,* species that do not live in permanent burrows. However, the author noted differences between burrows of *A. longa* according to their position within the soil profile. Burrows that opened to the surface were without a mucus lining whereas those deeper in the soil profile were lined with mucus.

Although the role of mucus in altering soil structure is not clear, it is generally accepted that the mucus lining aids the earthworm in movement through the soil, excretion, and respiration. According to Richards (1978), three types of mucus producing cells are found in most lumbricids. One cell type produces a mucopolysac-charide-protein-lipid complex that acts as a lubricant. A second type of cell produces a carboxylated acid mucus that is trapped by small projections of the cuticle and is important in forming a moist surface film for respiration exchange. The third type of cell forms a protein-rich secretion that may affect the viscosity and water-retention capabilities of the substance produced by the other gland cells.

As discussed earlier, burrows may be of a permanent or temporary nature and may or may not open to the surface depending upon species. Generally, burrows formed by anecique species open to the surface while those formed by endogé species do not open to the surface. Not only is the continuity of the burrows species dependent but continuity of the burrows from the surface of the soil to lower soil layers may be affected by tillage (Edwards et al., 1979). They reported large *L. terrestris* burrows in the subsoil that earthworms had abandoned for at least twenty-four years.

E. Burrow Function

1. Infiltration

For preferential flow of water to occur in soil, the soil pores must be continuous through a distance and have a source of low tension or tension-free incoming water. Not all large pores in the soil matrix contribute to the bypass flow of water. Nevertheless, water infiltration is two to ten times faster in soils with earthworms than in soil without earthworms (Lee, 1985). Factors such as tunnel continuity, diameter, length, casting placement, microrelief around the burrow opening, surface sealing, and whether the burrow is occupied by an earthworm should be considered when relating infiltration to earthworm tunnels.

Zachmann et al. (1987) measured macroporous infiltration and redistribution as affected by *Aporrectodea tuberculata* and *Lumbricus rubellus*. Infiltration rates and bromide movement were compared in tilled plots and in no-till plots, both with and without residue. Zachmann et al. (1987) reported increased burrowing activity by both species and increased infiltration rates in the treatments with residue. Infiltration and bromide movement in the upper 5 cm was faster in the tillage treatments with earthworms and residue. However, bromide movement beyond 19 cm was greatest in no-till with worms and residue. They concluded that the continuous macropores in the no-till treatments were responsible for the considerable movement of water and bromide through the 19 cm profile as compared to the tilled treatments.

For water to flow through macropores, the opening of the macropores must be in contact with tension-free water. Ehlers (1975) and Smittem and Collis-George (1986) measured the rate of infiltration into earthworm burrows and correlated the rate of infiltration with burrow radius. However, Edwards et al. (1979) modelled infiltration rates as a function of burrow diameter and length. They concluded that runoff was affected by both burrow diameter and length. Smittem (1986) measured water flow through a water-filled burrow into surrounding unsaturated soil and demonstrated that pore length was more highly correlated with flow rate than pore radius. Lee and Foster (1991) suggested that the importance of the burrow length was due to the surface of the burrow wall that allowed for lateral movement of the water in a water-filled burrow to infiltrate into the surrounding soil.

Surface seals may also alter infiltration rates. Interactions between earthworm activity and surface seals have not been fully investigated. Lal (1976) observed that earthworm activity in no-tilled plots reduced surface sealing and would help maintain high infiltration rates under these conditions. The importance of earthworms in reducing surface crusting was also demonstrated by Kladivko et al. (1986) in a greenhouse experiment with *L. rubellus*. In their study, surface crusting from simulated rainfall was reduced in pots containing earthworms compared to pots without earthworms. They concluded that the presence of an active earthworm population could be important in the reduction of surface crusting in cultivated soils. Similar conclusions were reported by Freebairn (1989) and Freebairn et al. (1989).

The influence of earthworm surface castings on rates of infiltration is unknown. Earthworms commonly leave castings near macropore openings. If these castings alter flow patterns around the macropores, infiltration rates may be reduced. Earthworms may also seal their burrows during rainfall and the burrows may have little effect on infiltration. In experiments with soil surface crust, Roth and Joschko (1991) reported that *Aporrectodea caliginosa* and *A. rosea* sealed their burrows (4-5 mm in diameter) within the first 15 minutes of a 30 mm/h simulated rainfall. Burrows of more than 5 mm diameter remained unsealed thus increasing infiltration rates.

F. Casts

Earthworm casts consist of excreted masses of mineral soil, which may be mixed with comminuted and partially digested plant residues. The shape, size, and composition of the casts are usually species-specific. Generally, small earthworms produce small casts, with a finer structure than the larger species (Lee, 1985). Most anecic species of earthworms, when they are forming burrows, ingest soil and eject it at the soil surface. Endogenic species, however, may form surface casts or deposit the waste material along the burrow walls or in subsurface cavities.

Surface casts have different shapes, sizes, structures, and compositions; therefore their significance in soil structure may differ. Lee (1985) recognized two basic forms of casts. One type of casts is characterized by paste-like slurries that form generally rounded but less regular shapes. The other type of casts is ovoidal, or subspherical to spherical pellets, ranging in size from less than 1 mm to greater than 1 cm, according to the species that formed them. Soil structure and resistance to erosion are both affected by the form of the casts. Globular casts are resistant structures and may persist for months, while granular casts may be washed away by rain and contribute to sheet erosion (Lavelle, 1988).

Quantitative data on the shapes of casts produced by earthworms is limited. McKenzie and Dexter (1987) compared the shapes of casts produced by *A. rosea* to that of similar sized aggregates from the soil. They digitized the centroid, area, and perimeter of casts produced by *A. rosea* and compared them to shapes of similar sized soil aggregates. Radius and curvature spectra were then calculated and it was demonstrated that earthworm casts were more round than soil aggregates of similar size.

Annual production of surface casts by earthworms may be as high as 20-30 kg per square meter (Graff, 1971; Lavelle, 1978). More commonly, cast production is 1-5 kg per square meter (Edwards and Lofty, 1977; Lee, 1985).

Cast production is related to soil moisture and soil temperature (Sharpley and Syers, 1977). In New Zealand, surface casting occurs primarily during the period from April to September and peak activity occurs in early June. Lack of soil moisture limits earthworm activity prior to May, and lower soil temperatures reduce activity as indicated by surface casts in late June and July.

Earthworm casting activity has also been shown to be affected by food sources. Shipitalo et al. (1988) investigated the effect of diet on the feeding and casting activity of *L. terrestris* and *L. rubellus* in laboratory cultures. In their studies, both species were provided with alfalfa, bromegrass, corn, red clover or no-food diets. Although *L. rubellus* has been reported not to deposit casts on the surface (Guild, 1955; Edwards and Lofty, 1977; Reynolds, 1977), in the study by Shipitalo et al. (1988) both species produced surface casts. Food consumption ranged from 5 to 52 mg per gm live worm per day for *L. rubellus* and 2-13 mg per gm live worm per day for *L. terrestris* for all foods except bromegrass. Both species rejected bromegrass. Cast production was also higher for *L. rubellus* with values ranging from 80 to 460 mg per gm live worm per day as opposed to 70 to 180 mg per gm live worm per day for *L. terrestris*.

Production of casts may also be influenced by pH and calcium content of the soil. Springett and Syers (1984) measured production of casts by *L. caliginosa* when $CaCO_3$, $Ca(OH)_2$, $CaSO_4$, and $Ca(NO_3)_2$ were added to the soil. Soil pH increased with the addition of $CaCO_3$ and $Ca(OH)_2$ but remained constant with increasing amounts of $CaSO_4$ and $Ca(NO_3)_2$. Under these conditions, cast production increased with increasing pH and Ca levels but did not increase with increasing Ca when the soil pH was constant. Cast production decreased when pH was above 7.3 with $CaCO_3$ and above 6.7 with $Ca(OH)_2$. Springett and Syers (1984) cited previous evidence that suggested that high osmotic pressure of the soil solution from the additions of high levels of $Ca(NO_3)_2$ was responsible for the decrease of cast production.

Behavioral traits that govern surface casts are little understood. Scullion and Ramshaw (1988) investigated factors that affected surface casting behavior of *A. longa, L. rubellus, A. caliginosa,* and *A. chlorotica.* The amount of surface casting activity was reported to be influenced by species interactions and was not related to soil bulk densities. In glasshouse experiments utilizing single species cultures, the amount of cast material deposited on the surface varied among species with *A. chlorotica* < *L. rubellus* < *A. caliginosa* < *A. longa.* In field communities, *L. terrestris* was dominant to *A. caliginosa, L. rubellus,* and *A. chlorotica.* Scullion and Ramshaw's results suggested that a single species dominated surface casting activity. They concluded that surface casting was a function of species, stage of growth within a species, and species assemblies. It should also be noted that greater surface cast production and the dominant role in casting within populations appeared to be partly related to an increase in species abundance.

Variation in casting activity may occur within a species. Jefferson (1958) reported that *A. caliginosa* produced surface cast at one site but at another site, 120 km away, casting activity by *A. caliginosa* was below the surface. Similar results have been presented for *A. rosea* by Bolton and Phillipson (1976), who reported that this species deposited casts primarily below the surface. However, Thompson and Davies (1974) found that *A. rosea* was a prolific surface caster. Lee (1985) attributed these differences to soil compaction and that species that normally deposit casts in voids below the surface may produce surface casts in compacted soils.

Earthworm casts contain higher proportions of clay and silt and lower proportions of sand than the surrounding soil (Nye, 1955; Sharpley and Syers, 1976). Maximum particle size is related to the relative sizes of the earthworm. Bolton and Phillipson (1976) measured maximum particle sizes in the posterior gut of *A. rosea, A. calignosa* and *Octolasion cyaneum.* Particle sizes were ca 100 μm, 200 μm, and 500 μm, respectively, which was proportional to the relative diameter of the three species.

Mechanisms by which earthworms produce and stabilize soil aggregates are complex and poorly understood. Earthworm casts are usually more water stable than aggregates of identical soils. However, freshly deposited casts appear less stable than other soil aggregates and stability increases with time after deposition (Shipitalo and Protz, 1987).

Cast stability is affected by a number of factors including: age of cast, structural organization of the soil, food source of the earthworms, and microbial interactions. Edwards and Lofty (1977) cite several theories for increased cast stability. Early research suggested that plant remains that pass through earthworms reinforce and hold the aggregates together. An alternative theory suggests that internal secretions cement the soil particles together as they pass through the earthworm intestines. It has also been reported that increased stability is due to gums produced by bacteria or by the development of fungal hyphae (Edwards and Lofty, 1977).

Several processes may be important in altering the stability of the casts. In the process of ingestion of food and mineral soil by the earthworm, large amounts of mucus are extruded and mixed with the ingested material. This mass is then forced through the digestive system by muscle contractions (Edwards and Fletcher, 1992). Stability of the material may decrease by the breaking of bonds between the soil particles (Griffith and Jones, 1965; Utomo and Dexter, 1981). Thus, when fresh and wet casts are secreted they would be susceptible to dispersion. Casts that were allowed to dry and then rewetted would become more stable as a result of the formation of micro-aggregates caused by the bonding between the primary particles (Greacen, 1960).

Microbial growth also increases casts stability. Bacterial cells produce gel coats to which clay particles adhere forming stable micro-aggregates (Emerson et al., 1986). Growth of fungal hyphae on the outside of soil aggregates and may stabilize macro-aggregates (Tisdall and Oades, 1982). The stabilizing action of fungal hyphae tends to become stronger after 10-15 days (Aspiras et al., 1971; Metzger et al., 1987).

Dispersion of the casts may also be related to organization of the parent soils. Marinissen and Dexter (1990) investigated the mechanisms used to stabilize earthworm casts and artificial casts using *A. caliginosa*. They studied casts from two soils with a similar texture but different in the degree of development. Worm casts produced in the "older" soils were much more stable than those in the "young" soils as compared to the artificial casts.

In a series of studies with *L. terrestris* and *L. rubellus,* Shipitalo et al. (1988) and Shipitalo and Protz (1987) investigated the effects of diet on casts stability. Both species were fed alfalfa (*Medicago sativa* L.), bromegrass (*Bromus inermis* Leyss.), red clover (*Trifolium pratense* L.), corn (*Zea mays* L.) or no food. They reported that the presence of organic debris was necessary for the stabilization of earthworm casts. Casts produced by *L. rubellus* were more stable than those produced by *L. terrestris*. The dissimilarity between the two casts was attributed to differences in the amounts of incorporated organic matter. *L. rubellus* casts had greater amounts of incorporated organic matter than did those of *L. terrestris*.

III. Decomposition and Nutrient Release

Many organisms are responsible for decomposing organic matter that reaches the soil. Soil microorganisms are the major decomposing agents that break down plant and animal residues into usable nutrients for plants and other soil organisms. Not all plant materials are subject to immediate decomposition by soil microorganisms. Some are first partially broken up and reduced in size by other soil animals. Earthworms, by their feeding activities and moving of plant and animal debris, contribute to this process.

Earthworms derive their nutritional requirements by feeding on plant litter as well as a wide range of decaying organic substances. Earthworms play a key role in removing plant litter, dung, and other organic material from the surface of the soil and comminuitive processes within the soil matrix. Typically, surface feeders are represented by *L. terrestris*. Earthworms in this group are primarily responsible for the transportation of organic materials from the surface layer into the soil matrix. In the absence of these earthworms, surface mats may develop, locking up large quantities of nutrients. Earthworms that feed within the soil volume are represented by *A. trapezoides, A. tuberculata, A. turgida, A. rosea,* and *O. tyrtaeum.* Earthworms in this group feed on the partially decomposed residue and microbes associated with the decomposing material. This suggests that residue decomposition may be controlled through comminution and microbial population regulation. Earthworms that feed in the soil volume may be important in regulating microbial populations and thus the rate of residue decomposition. Lavelle (1988) discussed the food resources in the soil in relation to composition and earthworm utilization. Leaf litter was considered the best resource because it is high in assimilable carbohydrates and low in lignocellulose, and has a favorable distribution over time and space. Old litter which is highly decomposed loses a large part of the nutritive value as compared to fresh litter. Soil organic matter was considered a poor quality resource as it is dispersed in the soil mineral matrix and may include up to 75% large, complex, humic molecules which are bound to clay particles (Stout et al., 1981).

The amount of organic material earthworms consume is governed by the amount available rather than by the capacity to ingest it (Lee, 1985). Data of Franz and Leitenberger (1948) for *L. rubellus,* van Rhee (1963) for several species of Lumbricids, and Needham (1957) for *L. terrestris* indicate that earthworms consume about 27 mg dry weight of litter per gram wet weight of earthworms per day.

Information on the rate of residue breakdown that can be contributed to earthworms is limited. Most efforts have been directed towards showing that earthworms are important in breaking down residue over a predetermined period of time. Mackay and Kladivko (1985) investigated the rate of breakdown of soybean and corn residues in soil using *L. rubellus.* Greenhouse studies with various populations of *L. rubellus* (0, 250 m^{-2} and 500 m^{-2}) showed that after 36 days, 60% of the soybean residue was recovered in pots without earthworms as compared to 34% for pots containing earthworms. With maize, 85% of the residue was recovered in pots without worms as compared to 52% in pots containing worms. Similarly, Zachmann and Linden (1989) determined the effects on *L. rubellus* on corn residue breakdown. Corn residue in the presence of earthworms was degraded 30% faster than without earthworms. These studies demonstrate the importance of earthworms in residue decomposition but not the rate at which residue is degraded.

IV. Enchytraeidae

Enchytraeidae are a family of Oligochaetes that occur in terrestrial, littoral, and aquatic habitats. They are included in this review because they are often confused with earthworms and in other cases with nematodes. Enchytraeids (potworms) are pale-colored worms that range in size from 10-20 mm in length (Nakamura, 1979). They are not as well known as other Oligochaeta; however, over 600 species have now been recognized (Dash, 1990). Usually these worms are found in habitats that are acidic, have a high organic matter content, and are not drought stressed.

The role of Enchytraeids in soil formation and litter breakdown is not well understood. Kubiena (1953) observed that in some mineral soils enchytraeids form "micro-sponge" structures in which clay-humus complexes form water stable aggregates. Zachariae (1963) observed that enchytraeids fed on finely divided plant debris and feces of litter feeding Collembola. He concluded that feces produced by the enchytraeids form a considerable portion of the recent humus of forest soils. Thompson et al. (1990) made similar observations in France in forest, pasture, and cultivated soils. They compared macromorphological and micromorphological features related to agriculture and faunal activity in three loess-derived soils in France. In forest soils (Typic Hapludalf), enchytraeid faeces occurred in all horizons but were most abundant in the A horizon. Under these conditions, fecal pellets ranged from 50-250 μm in diameter and often filled the channels and chambers in the soil. They described the pellets as rugose and clustered in loose, continuous infillings of the channels and chambers. Fecal pellets in pasture soils (Mollic Hapludalf) and cultivated soils (Typic Hapludoll) were not as abundant as in the forest soils. These pellets occurred as dense continuous infilling of the channels. Thompson et al. concluded that the activity of the soil fauna (enchytraeids and earthworms) was significant in determining the soil morphology within this region.

Other authors have concluded that enchytraeids are more important in the decomposition process than in soil formation. Dash (1990) suggested that enchytraeids contribute to the decomposition process by (1) increasing substrate availability for other decomposer organisms by fragmentation and partial digestion of plant residues, (2) production of fecal pellets that are added to the soil thereby providing sites for microbial activity, and (3) encouraging a high level of activity by the microflora by grazing on the aging microflora.

Several attempts have been made to assess the role of enchytraeids in the decomposition process by studying their feeding habits and food requirements. Jegen (1920) observed that enchytraeids feed on plant remains and silica. Clark (1949) reported that in Australian forest soil, enchytraeids ingest fragmented plant remains and large quantities of fungal mycelium. In studies on the feeding habits of three species of enchytraeid: *Achaeta eiseni, Cognettia cognetti*, and *Marionina cambrensis* on fungal mycelium, it was shown that *A. eiseni* and *C. cognetti* selectively feed on fungi and that *M. cambrensis* did not show any selectivity in its feeding (O'Connor, 1957). This work was supported by studies of Dash and Cragg (1972) in which they concluded that enchytraeids are able to select among species of fungi.

Temperature and soil moisture have been found to be the primary environmental limiting factor for the enchytraeid worms (O,Connor, 1967; Dash and Cragg, 1972; Abrahamsen, 1972). Although numerous studies have been made in North America on the distribution of the enchytraeids, limited population studies have been conducted. Population studies have been made for a large range of habitats and geographic distributions in Finland, England, Canada, India, and Japan (Table 3). In a comparison study on the abundance of enchytraeids in three habitats in England,

Table 3. Densities for Enchytraeidae in different habitats

Habitat	Density (meter^{-3})	Source
Pasture	30,000-74,000	Nielson, 1955
Woodland	134,000	O'Connor, 1957
Grassland	37,000-200,000	Peachey, 1963
Woodland	2,000-22,000	Dash and Cragg, 1972
Sorghum/clover	935-4,108	House and Parmelee, 1985
Sorghum/rye	520-1,837	House and Parmelee, 1985

most were found in *Juncus* Moor as compared to *Nardus* grassland and bare peat (Peachey, 1963).

In Georgia, House and Parmelee (1985) compared populations of soil arthropods and earthworms from four cropping systems: no-tillage sorghum/rye, no-tillage sorghum/clover, conventional-tillage sorghum/rye, and conventional-tillage sorghum/clover. Their data showed that greater numbers of enchytraeids were found in the conventional-tilled plots as compared to the no-tillage plots.

Enchytraeid feces may be important in soil formation in some soil types and cropping systems (Zachariae, 1964 [quoted in Dash and Cragg, 1972]). For example, in studies comparing macromorphological and micromorphological features related to agriculture and faunal activity in three loess-derived soils in France, it was found that in forest soils (Typic Hapludalf), enchytraeid feces occurred in all horizons but were most abundant in the A horizon (Thompson et al., 1990). Fecal pellets varied from 50-250 μm in diameter and often filled the channels and chambers. Faece pellets were rugose and clustered in loose, continuous infillings of the channels and chambers. Fecal pellets in pasture soils (Mollic Hapludalf) and cultivated soils (Typic Hapludoll) were not as abundant as in the forest soils. Pellets in the pasture and cultivated soils occurred as dense, incomplete or loose, continuous infilling of the channels.

Enchytraeids primarily feed on algae, bacteria, and fungi and would therefore be considered secondary decomposers (Dash and Cragg, 1972). Dash (1990) summarized their role in decomposition as (1) providing food sources for other decomposers, following the digestive processes of the enchytraeids, (2) producing fecal material that is added to the soil to provide sites for microbial activity, and (3) promoting a high level of activity by the microflora, which they do in their grazing activity by preventing aging of the microflora.

V. Other Soil Arthropods

Although many species of insects and Arachnida inhabit soils and are important in residue decomposition and modifying soil physical characteristics, this review will be limited primarily to Acarina (mites), Collembola (springtails), Hymenoptera (ants), and Isoptera (termites).

Soil is a diverse habitat and is inhabited with a diverse and large arthropod community. Soil arthropods range in size from 200 μm, for mesostigmatid mites

(Krantz and Ainscough, 1990), to 250 mm for the giant phasmid *Palophus titan* (Metcalf et al., 1962). Feeding habits of soil arthropods are as divergent as their size. Soil arthropods may exist as free living predators in the soil, in litter, or on plants where they feed on other small invertebrates. Others are internal or external parasites, fungivores or detritivores. Soil arthropods inhabit litter layers (epedaphic), soil mineral layers (euedaphic), or occupy both habitats at different life stages (hemiedaphic).

A. Collembola — Mites

Collembola are primitive wingless insects and usually range in size from 0.2-9 mm. They are found in most soil habitats and may be classified as epedaphic or as euedaphic. Generally those that live in the litter layer (epedaphic) are larger than those found in the soil matrix (euedaphic). The euedaphic forms are usually smaller (<2 mm) and are generally worm-like in appearance. Collembola have a wide range of diets which includes decayed vegetation, carrion, feces and/or bacteria and fungi that are associated with these substrates (Christiansen, 1990). Along with the Acari (mites), the Collembola are usually the most abundant soil arthropods (Hale, 1967). It is generally accepted that Collembola contribute to the soil by ingesting organic matter and producing fecal pellets (Hale, 1967).

Although several orders of mites occur in the soil, only saprovores (for example, decomposers) will be mentioned in this review. These mites feed on decomposing litter fragments and/or fungi and bacteria. Generally they are restricted to movement within pre-existing voids in litter or soil. Like the Collembola, mites are not considered important in modifying soil structure, water relationships or gas exchange below the litter layer. Both groups, however, enhance the disappearance of plant litter (Edwards and Heath, 1963; Witkamp and Crossley, 1966).

VI. Insects

Soils are modified by insects while digging and burrowing in search of food, while growing and developing, or in responding to environmental stimuli. Insects transport organic and inorganic materials either actively or passively. Examples are the movement of soil particles, food, and the deposition of feces (active). They passively transport materials by altering water movement, formation of particulate and soluble materials which may then be transported by water and wind (Anderson, 1988).

A. Hymenoptera

Ants and termites usually dominate the soil insect biomass and are probably more important in affecting soil structure than are other insects (Lee and Foster, 1991). They are social insects and thus more active in small areas surrounding their nests. However, there may be many nesting sites, for example, Baxter and Hole (1967) reported 1,531 active or recently active mounds made by *Formica cinerea montana* in a prairie in Southwestern Wisconsin.

Soil structure is affected by ants and termites through the construction of mounds, nests, and gallery systems. Their role in modifying soil structure includes depositing subsoil on the surface, reducing soil bulk density, increasing concentrations of organic matter, influencing nutrient cycles, and impeding soil horizonation through pedoturbation (Hole, 1961).

Ant and termite burrows are constructed to provide nests, food storage, or chambers for the growth of fungi. Ant burrows are usually vertical and may extend to depths of several meters. In many cases, other burrows may radiate laterally. These lateral burrows may be used for storage of food or to house fungus combs (Lee and Foster, 1991). These burrows may encompass large areas, for example, Baxter and Hole (1967) estimated the volume of the burrows and chambers of an ant *Formica cinerea* to be 0.02 m^3.

Quantitative data on the effects of ant and termite burrows on soil structure and infiltration rates are limited. However, Lobry de Bruyn and Conacher (1990) reviewed the role of termites and ants in Australia on soil modification and infiltration. They concluded that ants and termites either increased infiltration by improving soil structure and porosity or decreased infiltration by producing compact surfaces which assist runoff and erosion.

VII. Research Needs

Additional research is needed in many areas of soil fauna agroecology. Previously, research by soil zoologists has focused on the effects of soil environments on the fauna not on the effects of the fauna on the soil. This was due in part to the lack of available techniques for observing the behaviors of soil animals and the results of their activities *in situ*. Before we can quantify the effects of the soil fauna on soil structure, redistribution of water, and organic matter transformations, techniques for quantifying faunal activity within the soil matrix must be developed. Studies are also needed to determine interactions among plants, microbes, and macrofauna. Traditionally, ecological studies in agroecosystems have been developed on the concept of single species without regard to interactions and associations among species. Soil ecologists have developed extensive data sets on soil invertebrates and their effects on the decomposition and mineralization processes at several sites in North America. These sites include the Jornada desert (Santos and Whitford, 1981; Steinberger et al., 1984), shortgrass prairie ecosystem (Anderson et al., 1981; Coleman et al., 1983; Ingham et al., 1985), and the deciduous forests of Coweeta (Crossley, 1977a,b). House and Parmelee (1985) reported "soil fauna of natural terrestrial ecosystems influence organic matter decomposition, and mineralization processes such as nutrient release rates" (Crossley, 1977b; Peterson and Luxton, 1982; Seastedt, 1984). These processes, however, have not been demonstrated for the soil fauna of agricultural systems. Therefore, the role of the soil fauna in agricultural systems must be explored before models can be developed for determining the role of the soil fauna in soil formation and nutrient cycling.

One of the major problems in conducting needed research is the lack of sampling techniques to quantify soil organisms. Difficulties in comparing population estimates of the soil fauna often arise because the research techniques include a variety of sampling strategies, a lack of detailed soil descriptions, and a lack of detailed macroclimatic data. Standardized procedures need to be developed and implemented for conducting population estimates in a wide range of habitats.

Some of the specific questions that must be addressed concerning the role of earthworms are: (1) what are the species composition and abundance of earthworms under various farming systems, (2) how do tillage and farm management systems influence the abundance of the different species of earthworms under different soil types and climatic conditions, (3) what is the effect of different pest management programs (for example, herbicides, insecticides, and fertilizers) on earthworm populations and biological activity, (4) what is the relationship between plant species and earthworm diversity, (5) how is growth and reproduction of earthworms related to cover crops, and (6) can earthworms be utilized to penetrate clay hardpans and compacted soils?

Several studies have shown that earthworms are important in residue decomposition and nutrient cycling. The mechanisms involved in this process are poorly understood. Again some of the questions that must be addressed are: (1) what is the role of the different species of earthworms in the decomposition of plant roots and surface plant debris, (2) how do earthworms affect residue distribution, (3) how is the life stage of the earthworms related to residue decomposition, and (4) what is the relationship among earthworms, insects, and microbes in organic residue decomposition?

Earthworms are important in infiltration, but the mechanisms in this process are poorly understood. Research is needed to answer the following questions: (1) how does the condition of the soil surface affect the opening and closing of macropores, (2) how does macropore morphology affect flow rates, (3) what is the influence of surface casts on infiltration, (4) what is the effect of tillage practices on earthworm macropores, (5) how do various factors such as species of earthworms, stage of earthworm growth, temperature, moisture, soil type, soil compaction, and depth of food placement affect formation and function of macropores, and (6) what is the effect of surface casts on rill erosion?

References

Abrahamsen, G. 1972. Ecological study of enchytraidae (Oligochaeta). *Pedobiologia* 13:6-15.

Anderson, J.M. 1988. Invertebrate-mediated transport processes in soils. 1. Interactions between invertebrates and microorganisms in organic-matter decomposition, energy flux, and nutrient cycling in ecosystems. *Agric. Ecosystems Environ.* 24:5-19.

Anderson, R.V., D.C. Coleman, and C.V. Cole. 1981. Effects of saprotrophic grazing on net mineralization. p. 201-216. In: F.E. Clark and T. Rosswall (eds.), Terrestrial Nitrogen Cycles, *Ecol. Bull.*, Stockholm.

Aspiras, R.B., O.N. Allen, G. Chesters, and R.F. Harris. 1971. Chemical and physical stability of microbially stabilized aggregates. *Soil Sci. Soc. Am. Proc.* 35:283-286.

Barnes, B.T. and F.B. Ellis. 1979. Effects of different methods of cultivation, direct drilling, and dispersal of straw residues on populations of earthworms. *J. Soil Sci.* 30:669-679.

Baxter, F.P. and F.D. Hole. 1967. Ant (*Formica cinerea*) pedoturbation in a prairie soil. *Soil Sci. Soc. Am. Proc.* 31:425-428.

Berry, E.C. and D.L. Karlen. 1993. Comparison of Alternate Farming Systems: II. Earthworm population density and species diversity. *J. Alternative Agriculture* 8:23-28.

Bolton, P. J. and J. Phillipson. 1976. Burrowing, feeding, egestion, and energy budgets of *Allolobophora rosea* (Savigny) (Lumbricidae). *Oecologia* 23:225-245.

Bostrom, U. 1987. Growth of earthworms (*Allolobophora caliginosa*) in soil mixed with either barley, lucerne or meadow fescue at various stages of decomposition. *Pedobiologia* 30:311-321.

Bostrom, U. and A. Lof-Holmin. 1986. Growth of earthworms (*Allolobophora caliginosa*) fed shoots and roots of barley, meadow fescue and lucerne. Studies in relation to particle size, protein, crude fiber content and toxicity. *Pedobiologia* 29:1-12.

Bouché, M.B. 1971. Relations entre les structures spatiales et fonctionelles de écosystés, illustrées par le role pédobiologique des vers de terre. p. 187-209. In: P. Pesson (ed.), *La Vie dans les Sols*. Gauthiers Villars, Paris.

Bouché, M.B. 1972. Lombriciens de France. *Ecologie et Systématique*. Ann. Zool 72:1-671.

Bouché, M.B. 1975. Action de la faune sur les états de la matiémes. In: G. Kilbertus, O. Reisinger, A. Mourey, and J.A. Cancela da Fonseca (eds.), *Biodégradation et Humification*. Pierron: Sarreguemines.

Bouché, M.B. 1977. Stratégies lombricidennes. *Biol. Bull.* 25:122-132.

Christiansen, K.A. 1990. Insecta: Collembola. p. 965-995. In: D.L. Dindal (ed.), *Soil Biology Guide*. John Wiley & Sons, New York.

Clark, D.P. 1949. Thesis (Unpub.). University of Syndey. (Quoted in O'Connor, 1979.)

Clutterbuck, B.J. and D.R. Hodgson. 1984. Direct drilling and shallow cultivation compared with ploughing for spring barley on a clay loam in northern England. *J. Agric. Sci.* 102:127-134.

Coleman, D.C., C.P. Reid, and C.V. Cole. 1983. Biological strategies of nutrient cycling in soil systems. *Adv. Ecol. Res.* 13:1-55.

Coleman, D.C., R.E. Ingham, J.F. McClellan, and J.A. Trofymow. 1984. Soil nutrient transformations in the rhizosphere via animal-microbial interactions. *In:* J.M. Anderson, A.D.M. Rayner, and D.W.H Walton (eds.), *Invertebrate-Microbial Interactions*. Cambridge Univ. Press, Cambridge, p. 35-58.

Cotton, D.C.F. and J.P. Curry. 1980. The effects of cattle and pig slurry fertilizers on earthworms (Oligochaeta, Lumbricidae) in grassland managed for silage production. *Pedobiologia* 20:181-188.

Crossley, D.A. 1977a. The role of terrestrial saprophagous arthropods in forest soils: Current status of concepts. p. 49-56. In: W.J. Mattson (ed.), *The Role of Arthropods in Forest Ecosystems*. Springer-Verlag, New York.

Crossley, D.A. 1977b. Oribatid mites and nutrient cycling. p. 71-86. In: D.L. Dindal (ed.), Biology of Oribatid Mites. Syracuse, NY, State University NY at Syracuse, Coll. Environ. Sci. For.

Darwin, C.R. 1881. *The formation of vegetable mould through the action of worms with observations on their habits.* John Murray and Co., London. 298 pp.

Dash, M.C. 1990. Oligochaeta: Enchytraeidae. p. 311-340. In: Daniel L. Dindal (ed.), *Soil Biology Guide*. Wiley, New York.

Dash, M.C. and J.B. Cragg. 1972. Ecology of Enchytraeidae (Oligochaeta) in Canadian Rocky Mountain Soils. *Pedobiologia* 12:323-335.

De St. Remy, E.A. and T.B. Daynard. 1982. Effects of tillage methods on earthworm populations in monoculture corn. *Can. J. Soil Sci.* 62:699-703.

Dexter, A.R. 1978. Tunnelling in soil by earthworms. *Soil Biol. Biochem.* 10:447-449.

Edwards, C.A. 1983. Earthworm ecology in cultivated soils. p. 123-138. In: J.E. Satchell (ed.), *Earthworm Ecology from Darwin to Vermiculture*. Chapman, New York.

Edwards, C.A. and G.W. Heath. 1963. The role of soil animals in breakdown of leaf matter. In: J. Doeksen and J. van der Drift (eds.), *Soil Organisms*. North Holland, Amsterdam.

Edwards, C.A. and J.R. Lofty. 1977. *Biology of earthworms.* Chapman and Hall, London. 333 pp.

Edwards, C.A. and J.R. Lofty. 1982. The effect of direct drilling and minimal cultivation on earthworm populations. *J. Appl. Ecol.* 19:723-734.

Edwards, C.A. and K.E. Fletcher. 1992. Interactions between earthworms and microorganisms in organic-matter breakdown. *Agric., Ecosystems and Environ.* 24:235-247.

Edwards, W.M., R.R. Van der Ploeg, and W. Ehlers. 1979. A numerical study of the effects of noncapillary-sized pores upon infiltration. *Soil Sci. Soc. of Am. J.* 43:851-856.

Ehlers, W. 1975. Observations on earthworm channels and infiltration on tilled and untilled loess soil. *Soil Sci.* 119:242-249.

Eisenbeis, G. and W. Wichard. 1985. *Atlas on the Biology of Soil Arthropods.* Springer-Verlag, New York. 437 pp.

Emerson, W.W., R.C. Foster, and J.M. Oades. 1986. Organo-mineral complexes in relation to soil aggregation and structure. In: P.M. Huang and M. Schnitzer (eds.), Interactions of soil minerals with natural organics and microbes. *Soil Sci. Soc. Am. Spec. Publ.* 17:521-547.

Evans, A.C. 1948. Studies on relationships between earthworms and soil fertility. II. Some effects on earthworms on soil structure. *Ann. Appl. Biol.* 35:1-13.

Evans, A.C. and W.J. McL Guild. 1947a. Some notes on reproduction in British earthworms. *Ann. Mag, Nat. Hist.* 14:654-659.

Evans, A.C. and W.J. McL Guild. 1947b. Studies on the relationships between earthworms and soil fertility. I. Biological studies in the field. *Ann. Appl. Biol.* 34:307-330.

Evans, A.C. and W.J. McL Guild. 1948. Studies on the relationship between earthworms and soil fertility. IV. On the life cycles of some British Lumbricidae. *Ann. Appl. Biol.* 35:471-484.

Franz, H. and L. Leitenberger. 1948. Biologisch-chemische Untersuchungen uber Humusbildung durch Bodentiere. *Ost. Zool. Z.* 1:498-518.

Freebairn, D.M. 1989. Rainfall and tillage effects on infiltration of water into soil. Unpublished Ph.D. thesis. Univ. of Minnesota, St. Paul, MN.

Freebairn, D.M., S.C. Gupta, C.A. Onstad, and W.J. Rawls. 1989. Antecedent rainfall and tillage effects upon infiltration. *Soil Sci. Soc. Am. J.* 53:1183-1189.

Fuchs, D.J. and D.R. Linden. 1988. An earthworm population and activity survey of selected agronomic areas in Minnesota. Unpublished thesis, Univ. of Minnesota, St. Paul, MN. 27 pp.

Graff, O. 1953. Bodenzoologische Untersuchungen mit besonderer Berucksichtigung der terrikolen Oligochaeten. *Z. Pflanzern. Dung. Bodenk.* 61:72-77.

Graff, O. 1971. Stickstoff, Phosphor und Kalium in der Regenwurmlosung auf der Wiesenversuchsflache des Sollingprojektes. In: J.D'Aguilar (ed.). IV Colloquium Pedobiolgiae Dijon. *Institut National des Recherches Agriculturelles. Publ* 71-7, 503-511.

Greacen, E.L. 1960. Water content and soil strength. *J. Soil Sci.* 11:313-333.

Griffith, E. and D. Jones. 1965. Microbial aspects of soil structure: 1. Relationships between organic amendments, microbial colonization and changes in aggregate stability. *Plant and Soil* 23:17-33.

Guild, W.J. McL. 1955. Earthworms and soil structure. In: D.K. McE. Kevan (ed.), *Soil Zoology.* Butterworth, London.

Hale, W.G. 1967. Collembola. p. 397-412. In: A. Burges and F. Raw (eds.), Soil Biology. Academic Press, New York.

Harding, D.J.L. and R.A. Stuttard. 1974. Microarthropods. In: D.H. Dickinson and G.J.G. Pugh (eds.), *Biology of Plant Litter Decomposition.* Academic Press, London.

Hole, F.D. 1961. A classification of pedoturbations and some other processes and factors of soil formation in relation to isotropism and anisotropism. *Soil Sci.* 91:375-377.

Hole, F.D. 1981. Effects of animals on soil. *Geoderma* 25:75-112.

House, G.J. and B.R. Stinner. 1983. Arthropods in no-tillage soybean agroecosystems; community composition and ecosystem interaction. *Environ. Manage.* 7:23-28.

House, G.J. and R.W. Parmelee. 1985. Comparison of soil arthropods and earthworms from conventional and no-tillage agroecosystems. *Soil and Tillage Res.* 5:351-360.

Ingham, R.E., J.A. Trofymow, E.R. Ingham, and D.C. Coleman. 1985. Interactions of bacteria, fungi and their nematode grazers: effects on nutrient cycling and plant growth. *Ecol. Monogr.* 55:119-140.

Jeanson, C. 1961. Sur une methode d'etude du comortement de la faune du sol et de sa contribution a la pedogenese. C.R. Hebd. *Seance Acad. Sci.* 253:2571-2573.

Jeanson, C. 1964. Micromorphology and experimental soil zoology: contribution to the study, by means of giant-sized soil sections, of earthworm-produced artificial soil structure. p. 47-55. In: A. Jongerius (ed.), *Soil Micromorphology*. Proc. 2nd Int. Working Meeting Soil Micromorph., Arnhem.

Jeanson, C. 1971. Structure d'une galerie de lombric a la microsonde electronique. p. 513-525. In: J. D'Aguilar (ed.), *IV Colloquium Pedobiologiae. Institut National des Recherches Agronomiques Publ. No. 71-7,* Paris.

Jefferson, P. 1958. Studies on the earthworms of turf. C. Earthworms and casting. *J. Sports Turf Res. Inst.* 9:437-452.

Jegen, C. 1920. Bedeutung der Enchytraeiden fur die Humusbildung. *Landw. Jahrb. Schweiz.* 34:55-71.

Kemper, W.D., T.J. Trout, A. Segaeren, and M. Bullock. 1987. Worms and water. *J. Soil Water Cons.* XX:401-404.

Kladivko, E.J. and H.J. Timmenga. 1990. Earthworms and agricultural management. p. 192-216. In: J.E. Box and L.C. Hammond (eds.), *Rhizospere Dynamics*. American Association Advancement Sci.

Kladivko, E.J., A.D. Mackay, and J.M. Bradford. 1986. Earthworms as a factor in the reduction of soil crusting. *Soil Sci. Soc. Am. J.* 50:191-196.

Kobel-Lamparski, A. and F. Lamparski. 1983. Die Wiederbesiedlung flurbereinigten Rebgelandes durch Lumbriciden. Mitteilgn Dtsch Bodenkundl Gesellsch. 36:337-342.

Kobel-Lamparski, A. and F. Lamparski. 1987. Burrow constructions during the development of *Lumbricus badensis* individuals. *Biol. Fertil. Soils.* 3:125-129.

Krantz, W.W. and B.D. Ainscough. 1990. Acarina: Mesostigmata (Gamasida). p. 583-666. In: D.L. Dindal (ed.), *Soil Biology Guide*. John Wiley & Sons, Inc., New York.

Kretzschmar, A. 1978. Quantification écologique des galeries de lombriciens. Techniques et premiéres estimations. *Pedobiologia* 18:31-38.

Kretzschmar, A. 1982. Description des galeries des vers de terre et variation saisonnié des réseaux (observations en conditions naturelles). *Rev. Ecol. Bio. Sol.* 19:579-591.

Kubiena, W.L. 1953. *The Soils of Europe*. Thomas Murphy, London. 318 pp.

Lal, R. 1976. No-tillage effects on soil properties under different crops in western Nigeria. *Soil Sci. Soc. Am. J.* 40:762-768.

Lavelle, P. 1978. Les vers de terre de la savane de Lamto (Cote d'Ivoire). Peuplements, populations et fonctions de l'écosystéme. *Publ. Lab. Zool. E.N.S.* 12:1-301.

Lavelle, P. 1988. Earthworm activities and the soil system. *Biol. Fertil. Soils* 6:237-251.

Lee, K.E. 1959. The earthworm fauna of New Zealand. *N.Z. Ept. Sci. Industr. Res. Bull.* 130. 486 pp.

Lee, K.E. 1985. Earthworms: Their ecology and relationship with soils and land use. Academic Press, New York. 411 pp.

Lee, K.E. and R.C. Foster. 1991. Soil Fauna and Soil Structure. *Aust. J. Soil Res.* 29:745-775.

Lobry de Bruyn, L.A. and A.J. Conacher. 1990. The Role of Termites and Ants in Soil Modification: A Review. *Aust. J. Soil Res.* 28:55-93.

Lofs-Holmin, A. 1983. Earthworm population dynamics in different agricultural rotations. p. 151-160. In: J.E. Satchell (ed.), *Earthworm Ecology from Darwin to Vermiculture*. Chapman and Hall, New York.

Mackay, A.D. and E.J. Kladivko. 1985. Earthworms and rate of breakdown of soybean and maize residues in soil. *Soil Biol. Biochem.* 17:851-857.

Madge, D.S. 1969. Litter disappearance in forrest and savanna. *Pedobiologia* 9:288-299.

Maldague, M.E. 1970. Role des animaux édaphiques dans la fertilité des sols forestiers. Publ. Inst. Natinal Etude Agronomique Congo. *Sér. Sci.* 112:1-245.

Marinissen, J. and A.R. Dexter. 1990. Mechanisms of stabilisation of earthworm casts and artificial casts. *Biol. Fertil. Soils* 9:163-167.

Martin, N.A. 1982. The interaction between organic matter in soil and the burrowing activity of three species of earthworms (Oligochaeta: Lumbricidae). *Pedobiologia* 24:185-190.

McKenzie, B.M. and A.R. Dexter. 1987. Physical properties of casts of the earthworm *Aporrectodea rosea*. *Biol. Fertil. Soils* 5:152-157.

McKenzie, B.M. and A.R. Dexter. 1988a. Axial pressures generated by the earthworm *Aporrectodea rosea*. *Biol. Fertil. Soils* 5:323-327.

McKenzie, B.M. and A.R. Dexter. 1988b. Radial pressures generated by the earthworm *Aporrectodea rosea*. *Biol. Fertil. Soils* 5:328-332.

Metcalf, C.L., W.P. Flint, and R.L. Metcalf. 1962. *Destructive and Useful Insects*. McGraw-Hill, New York. 1087 pp.

Metzger, L.D. Levanon, and U. Mingelgrin. 1987. The effect of sewage sludge on soil structural stability: Microbial aspects. *Soil Sci. Soc. Am. J.* 51:346-351.

Nakamura, Y. 1979. Micro-distribution of Enchytraeidae in *Zoysia* grassland. *Bull. Natl. Grassl. Res. Inst.* No. 14:21-27.

Needham, A.E. 1957. Components of nitrogenous excreta in the earthworms *Lumbricus terrestris* L. and *Eisenia foetida* (*Savigny*). J. Exp. Biol. 34:425-446.

Neuhauser, E.F., D.L. Kaplan, M.R. Malecki, and R. Hartenstein. 1980. Materials supporting weight gain by the earthworm *Eisenia foetida* in waste conversion systems. *Agricultural Wastes* 2:43-60.

Nielson, C.O. 1955. Studies on Enchytraeidae. 5. Factors causing seasonal fluctuations in numbers. *Oikos* 6:153-169.

Nye, Ph.H. 1955. Some soil forming processes in the humid tropics. IV. The action of the soil fauna. *J. Soil Sci.* 6:73-83.

O'Connor, F.B. 1957. An ecological study of the enchytraid worm population of a coniferous forest soil. *Oikos* 8:161-191.

O'Connor, F.B. 1967. The Enchytraeide. p. 213-266. In: A. Burges and F. Raw (eds.), *Soil Biology*. Academic Press, New York.

Parkinson, D. 1988. Linkages between Resource Availability, Microorganisms and Soil Invertebrates. Agriculture, *Ecosystems and Environ.* 24:21-32.

Peachey, J.E. 1963. Studies on the Enchytraeidae (Olichaeta) of moorland soil. *Pedobiologia* 2:81-95.

Perel, T.S. 1977. Difference in lumbricid organization connected with ecological properties. *Ecol. Bull.* 25:56-63.

Persson, T., E. Baath, M. Clarholn, H. Lundkvist, B.E. Soderstrom, and B. Sohlenius. 1980. Trophic structure, biomass dynamics, and carbon metabolism of soil organisms in a scotch pine forest. *Ecological Bull.* 32:419-459.

Peterson, H. and M. Luxton. 1982. A comparative analysis of soil fauna populations and their role in decomposition processes. *Oikos* 39:287-388.

Piearce, T.G. 1978. Gut content of some lumbricid earthworms. *Pedobiologia* 18:153-157.

Reiners, W.A. 1973. Terrestrial detritus and the carbon cycle. p. 303-327. In: G.M. Woodwell and E.V. Pecan (eds.), Carbon in the biosphere. *24th Brookhaven Symp. Biol.* Springfield, VA.

Reynolds, J.W. 1970. The relationship of earthworm (Oligochaeta: Lumbricidae and Megascolecidae) distribution and biomass to soil type in forest and grassland habitats at Oak Ridge National Laboratory. *Assoc. Southeast Bio. Bull.* 17:60.

Reynolds, J.W. 1972. The relationship of earthworm (Oligochaeta: Acanthodrilidae and Lumbricidae) distribution and biomass in sex heterogeneous woodlot sites in Tippecanoe County, Indiana. *J. Tenn. Acad. Sci.* 47:63-67.

Reynolds, J.W. 1977. The earthworms (Lumbricidae and Sparganophilidae) of Ontario. The Royal Ontario Museum, Canada. 141 pp.

Rhee, J.A. van. 1963. Earthworm activities and the breakdown of organic matter in agricultural soils. In: J. Doeksen and J. van der Drift (eds.), *Soil Organisms.* North Holland Publishing Co., Amsterdam.

Rhee, J.A. van and S. Nathans. 1961. Observations on earthworm populations in orchard soils. *Neth. J. Agric. Sci.* 9:94-100.

Richards, K.S. 1978. Epidermis and cuticle. In: P.J. Mill (ed.), *Physiology of Annelids.* Academic Press, London. 683 pp.

Roth, C.H. and M. Joschko. 1991. A note on the reduction of runoff from crusted soils by earthworm burrows and artificial channel. *Z. Pflanzenernahr. Bodenk.* 154:101-105.

Santos, P.F. and W.G. Whitford. 1981. The effects of microarthropods on litter decomposition in a Chihuahuan desert ecosystem. *Ecology* 62:654-663.

Satchell, J.E. 1967. Lumbricidae. p. 259-322. In: A. Burges, and F. Raw (eds.), *Soil Biology.* Academic Press, London.

Satchell, J.E. 1983. Earthworm ecology in forest soil. p. 161-170. In: J.E. Satchell (ed.), *Earthworm Ecology: from Darwin to Vermiculite.* Chapman and Hall, London.

Scheu, S. 1987. The role of substrate feeding earthworms (Lumbricidae) for bioturbation in a beechwood soil. *Oecologia* 2:192-196.

Scullion, J. and G.A. Ramshaw. 1988. Factors affecting surface casting behavior in several species of earthworm. *Biol. Fertil. Soils* 7:39-45.

Seastedt, T.R. 1984. The role of microarthropods in decomposition and mineralization processes. *Annu. Rev. Entomol.* 29:25-46.

Seastedt, T.R. and D.A. Crossley. 1984. The influence of arthropods on ecosystems. *BioScience* 34:157-161.

Sharpley, A.N. and J.K. Syers. 1976. Potential role of earthworm casts for the phosphorus enrichment of run-off waters. *Soil Biol. Biochem.* 8:341-346.

Sharpley, A.N. and J.K. Syers. 1977. Seasonal variation in casting activity and in the amounts and release to solution of phosphorus forms in earthworm casts. *Soil Biol. Biochem.* 9:227-231.

Shipitalo, M.J. and R. Protz. 1987. Comparison of morphology and porosity of a soil under conventional and zero tillage. *Can. J. Soil Sci.* 67:445-456.

Shipitalo, M.J., R. Protz, and A.D. Tomlin. 1988. Effect of diet on the feeding and casting activity of *Lumbricus terrestris* and *L. rubellus* in laboratory culture. *Soil Bio. Biochem.* 20:233-237.

Slater, C.S. and H. Hopp. 1947. Relation of fall protection to earthworm populations and soil physical conditions. *Soil Sci. Soc. Am. Proc.* 12:508-511.

Smittem, K.R.J. 1986. Analysis of water flow from cylindrical macropores. *Soil Sci. Soc. Am. J.* 50:1139-1142.

Smittem, K.R.J. and N. Collis-George. 1986. The influence of cylindrical macropores on steady-state infiltration in a soil under pasture. *J. Hydrol.* 79:107-114.

Springett, J.A. and J.K. Syers. 1984. Effect of pH and calcium content of soil on earthworm cast production in the laboratory. *Soil Biol. Biochem.* 16:185-189.

Steinberger, Y., D.W. Freckman, L.W. Parker, and W.G. Whitford. 1984. Effects of simulated rainfall and litter quantities on desert soil biota: nematodes and microarthropods. *Pedobiologia* 26:267-274.

Stout, J.D., K.M. Goh, and T.A. Rafter. 1981. Chemistry and turnover of naturally occurring resistant organic compounds in soil. In: E.A. Paul and J.N. Ladd (eds.), *Soil Biochemistry*, Dekker, New York.

Swift, J.J., O.W. Heal, and J.M. Anderson. 1979. Decomposition in terrestrial ecosystems. *Studies in Ecology.* Vol. 5, Blackwell Scientific Publications, London. 372 pp.

Thompson, A.J. and D.M. Davies. 1974. Production of surface casts by the earthworm *Eisenia rosea. Can. J. Zool.* 52:659.

Thompson, M.L., N. Fedoroff, and B. Fournier. 1990. Morphological features related to agriculture and faunal activity in three loess-derived soils in France. *Geoderma* 46:329-349.

Tisdall, J.M. 1978. Ecology of earthworms in irrigated orchards. p. 297-308. In: W.W. Emerson, R.D. Bond and A.R. Dexter (eds.) *Modification of Soil Structure.* Wiley, Chichester,

Tisdall, J.M. and J.M. Oades. 1982. Organic matter and water stable aggregates in soils. *J. Soil Sci.* 33:141-163.

Tomlin, A.D. and J.J. Miller. 1988. Impact of ring-billed gull (*Larus delawarensis* Ord.) foraging on earthworm populations of Southwestern Ontario agricultural soils. *Agric. Ecosystems and Environ.* 20:165-173.

Utomo, W.H. and A.R. Dexter. 1981. Age hardening of agricultural top soils. *J. Soil Sci.* 32:335-350.

Visser, S. 1985. Role of soil invertebrates in determining the composition of soil microbial communities. In. E.A.H. Fitter (ed.), *Ecological Interactions in soil.* Blackwell Scientific Publ., London. 451 pp.

Witkamp, M. and D.A. Crossley. 1966. The role of microarthropods and microflora in breakdown of white oak litter. *Pedobiologia* 6:293-303.

Zachariae, G. 1963. Was leisten Collembolen fur den Waldhumus? In: J. Van der Drift and J. Doeksen (eds.), *Soil Organisms,* Amsterdam.

Zachariae, G. 1964. Welche Bedeutung haben Enchytraeen im Waldboden. p. 57-68. In: A. Jongerius (ed.), *Soil Micromorphology.* Elsevier Pub. Co., Amsterdam.

Zachmann, J.E. and D.R. Linden. 1989. Earthworm effects on corn residue breakdown and infiltration. *Soil Sci. Am. J.* 53:1846-1849.

Zachmann, J.E., D.R. Linden, and C.E. Clapp. 1987. Macroporous infiltration and redistribution as affected by earthworms, tillage, and residue. *Soil Sci. Soc. Am. J.* 51:1580-1586.

Cycling of Nitrogen through Microbial Activity

J.L. Smith

I.   Introduction

Terrestrial scientists are generally in agreement that nitrogen (N) is the most limiting nutrient for plant growth. In most ecosystems the total N content of the soil and litter is greater than the total N content of the vegetation. Since on a global basis over 90% of the yearly N flux is between the soil and vegetation (Rosswall, 1976) it is obvious that microbial transformation of N is a major factor controlling plant growth. On an ecosystem basis soil N is generally not related to total plant biomass carbon (C) or net primary production but is correlated with soil C (Smith and Paul, 1990). The soil N concentration to a 1 m depth ranges from 0.02% in a desert scrub ecosystem to 0.24% in swamp and marshland. Other ecosystems, concentrations are 0.03% for savanna and woodland and 0.07% for cultivated land and tropical forests.

Between 90 to 95% of the total soil N is associated or combined with the soil organic fraction. This organic fraction generally termed soil organic matter (SOM) also contains approximately 40% of soil phosphorus (P) and 90% of soil sulfur (S) thus comprising a significant source of plant macronutrients (Smith et al., 1992). Concentrations of SOM range from 0.2% in desert soils to over 80% in peat soils. In temperate regions SOM ranges between 0.4 and 10%, with humid soils averaging 3-4% and semi-arid soils 1-3% (Smith and Elliott, 1990). Even in soils with relatively low concentrations of SOM this complex substance has a major influence on soil properties. SOM influences soil aggregation which in turn controls air and water relationships for plant root growth. Aggregation

This chapter was prepared by a U.S. government employee as part
of his official duties and legally cannot be copyrighted.

and SOM concentration control water infiltration and storage and provide resistance to water and wind erosion. The concentration of SOM also dictates the level of the microbial population, the so-called carrying capacity, and it's activity. It is this interactive relationship between SOM and the microbial population that controls nutrient cycling in most ecosystems.

The soil microbial population consisting of bacteria, fungi and microfauna are termed the soil microbial biomass (SMB). This soil pool is closely related to the SOM pool (Jenkinson and Ladd, 1981; Anderson and Domsch, 1989; Smith and Paul, 1990) and is measured as the amount of C and N in the SMB thus the terms SMB-C and SMB-N. SMB plays a major role as a catalyst in the decomposition of SOM and the release of inorganic nutrients in to the bulk soil where they become available for plant uptake. During decomposition the SMB assimilates complex organic substrates for energy and biomass carbon (C) with excess inorganic nutrients being released to the soil solution. This decomposition and nutrient release process is controlled by temperature, moisture, and the magnitude and quality of SOM as a microbial substrate. Perturbations such as soil wetting and drying, soil disturbance and organic substrate inputs will also influence the extent and rate of SOM decomposition. In soil systems microbially mediated decomposition and transformation of SOM is the primary driving force in nutrient cycling which plays a significant role in ecosystem development and functioning.

In the N cycle the SOM-N pool is considered to be the major source supplying the inorganic N pool, however the SMB-N pool my also contribute significantly to the inorganic N pool even though this pool of N may be small compared to the SOM pool. In most ecosystems the yearly N input is low and only comes from atmospheric wet and dry deposition and N fixation. Thus plant productivity in these ecosystems is dependent upon the uptake of soil inorganic N which is produced through microbial turnover and/or mineralization processes. Even in fertilized cultivated cropland 50% or more of the crop N requirement may come from the mineralization of SOM-N. Since the turnover of SOM is slow compared to the turnover of SMB and the C/N ratio of SMB is much lower than that of SOM it can be hypothesized that the majority of the N for plant uptake is actually being released from the SMB pool. This hypothesis could be of importance to disturbed ecosystems such that by manipulating and managing the SMB disturbed systems may become more productive and/or recover more quickly. The ultimate question being whether man can manage to our benefit the SMB in a natural or disturbed ecosystem.

In this chapter I will discuss microbial biomass, it's measurement and activity in relationship to N cycling. Examples suggesting that a significant portion of plant available N comes from the turnover of the SMB will be presented.

II. Nitrogen in Terrestrial Soils

On a global basis the living biota and soil contain 332 Gt-N $(Gt=10^{15}g)$ compared to 2×10^8 Gt-N for rocks, sediments and coal (Soderlund and Svensson, 1976). For the terrestrial system 13 Gt-N is sequestered by the plant biomass, 0.2 Gt-N by animals, 2.0 Gt-N by litter (Soderlund and Svensson, 1976) and 88 Gt-N by soil (Zinke et al., 1984). Table 1 gives soil C and N values for 12 ecosystem types and global totals. Boreal forest, tundra and marshes represent a small amount of land area (18%) but contain 37% of the total soil C and 31% of the N. Temperate climate soils make up 28% of the total land area and contain 24 and 31% of the total soil C and N respectively. The global soil C/N ratio is approximately 15:1 based on a 1 m depth. The only N additions to natural ecosystems comes from atmospheric deposition and biological N fixation. The total input from these sources has been estimated at 0.20 Gt-N/yr, 0.14 Gt/yr from biological N fixation. In contrast the loss of N from terrestrial systems is estimated to range between 0.15 and 0.27 GT-N/yr, mainly due to biological denitrification.

On a local scale N is concentrated in the upper 10 to 15 cm of the soil profile. The low and uniform percent N values of the several ecosystems previously mentioned reflect the sampling depth of 1 m, most surface soils range from 0.05 to 0.3% N depending upon the ecosystem. Soil N generally declines exponentially with depth and can decrease by an order of magnitude at 1 m depth. The exponential decrease of N with depth mirrors that of SOM and SMB thus these organic pools of C and N are related and vary together. Most surface soils have a low concentration of inorganic N (NH_4 and NO_3) present at any one time, usually less than 1% of the total soil N. Due to these conditions most N cycling studies concentrate on the 0-15 cm zone when measuring processes such as mineralization and immobilization of N.

In most ecosystems N is concentrated in the SOM pool with little in the plant biomass. Figure 1 shows the N distribution in an annual grassland system. Most of the non-SOM N is contained in the microbial biomass, with equal amounts in the live plants and detritus. For this Mediterranean grassland ecosystem we also found that the flux of N through the SMB was equal to the yearly N uptake by the grass. On a proportion basis other ecosystems show similar N distributions with main differences be the yearly flux rates from various N pools (Rosswall, 1976).

The factors which affect soil N content are the soil forming factors described by Jenny (1941) which are; climate, topography, vegetation, parent material, and time. Since climate controls plant species and growth and governs microbial activity it is the most predominate factor affecting soil N, although the cultivation of soil has even more drastic long-term effects on soil N .

Over time N accumulates with succession to equilibrium levels at climax vegetation. It has been shown that young landscapes such as after volcanic eruptions can accumulate N rapidly after recolonization by N fixing plants (Halvorson et al., 1991). Parent material affects texture which can in turn give

Table 1. Total carbon and nitrogen content of terrestrial soil systems

Ecosystem type	Area (10^{12} m^2)	Carbon[a] density (kg m^{-2})	Nitrogen[a] density (kg m^{-2})	Soil carbon (Gt)	Soil nitrogen (Gt)
Tropical rain forest	17.0	15.3	0.76	260	12.9
Tropical seasonal forest	7.5	10.6	0.94	80	7.0
Temperate evergreen forest	5.0	12.7	0.78	64	3.9
Temperate deciduous forest	7.0	7.1	0.66	50	4.6
Boreal forest	12.0	15.5	1.10	186	13.2
Woodland and shrubland	8.0	5.4	0.32	43	2.6
Savanna	15.0	5.4	0.32	81	4.8
Temperate grassland	9.0	10.5	0.79	95	7.1
Tundra and alpine	8.0	21.8	1.15	174	9.2
Desert scrub	18.0	3.3	0.26	59	4.7
Cultivated land	14.0	7.9	0.84	111	11.8
Swamp and marsh	2.0	72.3	2.90	145	5.8
Totals	122.5			1348	87.6

[a] To a depth of 1 m.
Compiled by Smith and Paul, 1990; from Whittaker and Likens, 1973 and Zinke et al., 1984.

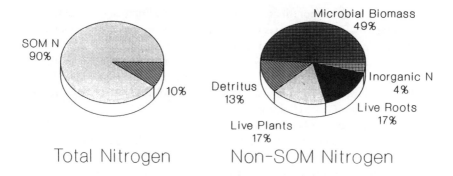

Figure 1. Typical N distribution for an annual grassland soil.

protection to SOM by adsorption to clays preventing decomposition by the SMB. In addition finer textured soils have greater SMB populations which will sequester greater amounts of soil N. Topography may affect the soil N level by modification of the local microclimate, where a ridge-top may be more arid than a bottom slope area thus affecting plant organic matter production.

Climate plays a duel role in affecting soil N content. Soil N concentration declines with increasing mean annual temperature over a wide range of soils and climates (Stevenson, 1982). Presumably due to more rapid decomposition of stabilized humus relative to the increased plant biomass production. Precipitation also affects SOM-N by increasing plant biomass which contributes to greater SOM levels and N containing humus. It is assumed that humus is a stabilizing factor for soil N thus ecosystems with greater humus synthesis will have a greater soil N concentration. It has been proposed that grasslands have greater SOM concentrations than forests because there is more humus synthesis in the rhizosphere of grasslands (Stevenson, 1982). It is the vegetation factor that separates ecosystems on the basis of SOM-N.

In recent years attention has again shifted to the fate of N in the environment. This is due to concerns of N pollution from NO_3 leaching into groundwater, gaseous loss of N such as N_2O affecting the atmospheric ozone and declining soil quality due to cultivation and loss of soil N. In the U.S. regulations are being implemented to limit N fertilization in areas with contaminated groundwater or shallow water tables. Recently studies of vadose zone N have shown that denitrification may be an important process in protecting groundwater (Geyer et al.,1992). Our challenge is to tighten the N cycle of agricultural ecosystems to maintain high productivity with environmental compatibility.

III. Microbial Biomass

The microbial biomass is the driving force behind SOM transformations and nutrient cycling in soil systems. Through the metabolism of complex SOM substrates the biomass contributes significantly to plant nutrition and ecosystem functioning. In the last 15 years there has been an increasing interest in measuring the SMB, mainly to compare treatment effects or make cross ecosystem comparisons. However by quantitatively measuring SMB it is possible to describe microbial mediated processes, such as N mineralization, on a mechanistic basis. Using SMB as a quantitative pool can also enhance models of N cycling (see Groot et al., 1991).

Several methods are available to quantitatively measure the SMB pool, the most widely used being the chloroform fumigation incubation method (CFIM) developed by Jenkinson and Powlson (1976). Other methods such as the substrate induced respiration (SIR) method (Anderson and Domsch, 1978) and the fumigation direct extraction (FDE) method (Brookes et al., 1985; Sparling and West, 1988) have been developed for more rapid analysis of SMB. Each of these methods are useful for measuring SMB and both the fumigation techniques can be used to measure SMB-N as well. Details of the CFIM and SIR are given in Parkinson and Paul (1982). Calibration of the FDE for estimating SMB-C is discussed by Tate et al.(1988) and Gregorich et al.(1990).

The advantages of the CFIM are that it is widely accepted, can measure both SMB-C and N, uses simple laboratory equipment and can be easily be modified for different soils or experiments. The disadvantages are the debate over a proper control value (discussed below), the long incubation time (10 d) and the occurrence of mineralization/immobilization reactions during the incubation period.

The advantages of the SIR method are that it has been calibrated with the CFIM, it is a rapid procedure using a small sample and a control sample is not needed. Since the method uses a substrate addition for maximum respiration a substrate saturation curve is needed for each soil. SMB-N cannot be measured using the SIR method and a gas chromatograph is essential.

The advantages of the FDE method are the short analysis time, it avoids mineralization/immobilization reactions, and provides good results with acidic soils for both SMB-C and N. The disadvantages are the less sensitive titration method for C analysis, the k value not being standardized and the length of fumigation being critical.

Using the CFIM the SMB-C value is calculated as

$$C_f - C_c / k_c \qquad\qquad (1)$$

where C_f is the CO_2-C from the fumigated sample, C_c the CO_2-C from a non-fumigated sample and k_c the proportion of organism C mineralized to CO_2 (see Voroney and Paul,1984).

Using the control C_c assumes that SOM is mineralized to the same extent in each system thus the subtraction of background C before SMB is calculated.

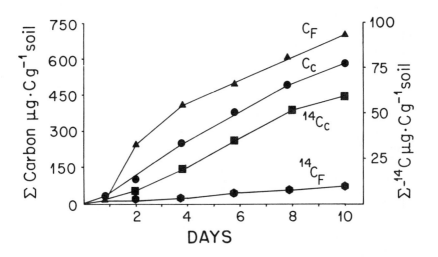

Figure 2. Total C evolved and ^{14}C labelled C evolved from soil where ^{14}C labelled leached straw was added and half the soil fumigated (C_f, $^{14}C_f$) and half used as a nonfumigated control (C_c, $^{14}C_c$).

Figure 2 shows data from an experiment where we added leached ^{14}C labeled straw to soil, to mimic native SOM, then measured SMB using the CFIM. The data clearly indicate that more $^{14}CO_2$ was evolved from the control sample than the fumigated sample. Using the control value C_c would underestimate the actual SMB as calculated by Equation 1. A more appropriate control can be calculated using the ratio of ^{14}C labeled straw evolved from each sample

$$C_{cc} = \frac{(^{14}C_f)}{(^{14}C_c)} * C_c \qquad (2)$$

where Ccc is the corrected control C to be used in the SMB-C calculation, ^{14}C from both samples and Cc is the total C evolved from the control sample. Table 2 shows three methods of calculating SMB-C based on different controls for different ecosystems. The SMB-C is calculated subtracting (1) no control value, (2) the total CO_2 evolved from the control sample and (3) a corrected control value. In this example one could argue that each method could be correct since the true value is unknown, however, in many instances subtracting the total control value results in negative SMB-C values. In addition Figure 2 and data from numerous other soils not presented clearly indicate that the control value to be used should be greater than zero but less than the total control CO_2-C.

Other factors affecting SMB measurements are related to pretreatment such as sieving, air-drying and storage (Ross et al., 1980; Ocio and Brookes, 1990),

Table 2. Correction values, CO_2-C flush and microbial biomass calculations for three diverse ecosystems

Ecosystem	CO_2-C Evolved		Correction[a] factor	Microbial biomass[b]		
	Fumigated	Control		C_f/k_c	C_f-C_c/k_c	C_f-C_{cc}/k_c
	µg C/g soil		%	µg C/g soil		
Grassland	783	595	16	1910	458	1678
Forest	265	234	20	646	76	532
Agricultural	234	172	12	571	151	520

[a] $= {}^{14}C_f/{}^{14}C_c \times 100$.

[b] $= C_f$: carbon from fumigated soil, C_c: carbon from control sample, C_{cc}: corrected carbon control, k_c: 0.41.

From Smith and Paul, 1986.

moisture content (Ross, 1989; Ritz and Wheatley, 1989) and organic matter content (Schnurer et al., 1985). The processes of air-drying and sieving tend to release organic C into the system which can affect the control in CFIM, theglucose addition in SIR and the fumigation process in CFIM and FDE. Wetting up and preincubation of a soil sample will affect the SMB population and the fumigation procedure. Problems of measuring SMB in acid soils and soils with recent additions of organic material have been noted (Jenkinson, 1988).

SMB values reported in the literature vary by greater than an order of magnitude ranging from 110 kg-C/ha in a cultivated field with total C of 0.7% to 2240 kg-C/ha in a grassland soil with total carbon of 7.0% (Smith and Paul, 1990). SMB-N values range from 40 to 400 kg N/ha based on 23 published reports from 13 countries. Global averages for SMB-C are 700, 1090, 850 for cultivated, grassland and forest soils respectively. For SMB-N the values are 195, 225 and 170 for cultivated, grassland and forest respectively. Thus the C/N ratio of the SMB ranges from 3.6 to 5.0 a lower value than expected and likely caused by the underestimation of SMB-C due to subtracting the entire control CO_2 when using CFIM.

IV. Microbial Activity and N Mineralization

Carbon dioxide is the major product of microbial metabolism of organic substrates for energy and biomass production. This C mineralization is accompanied by inorganic nutrient (N,P,S) release to the soil solution if the inorganic nutrient in the substrate is in excess of the microbial requirement for biomass synthesis. The SMB and it's activity are governed by the amount and metabolic availability of C, most of which is produced from plant growth. For the SMB it is C that is the most limiting element in most ecosystems. Figure 3 shows the energetic cycle of microbial biomass at a steady-state condition. In this scheme there are two pools of SMB, one active and one non-active or sustaining, both utilizing the same organic substrates. The active SMB pool releases C from growth and maintenance activities and the sustaining pool only from maintenance. The inorganic N pool fluctuates with mineralization/ immobilization processes. It has been hypothesized that the SMB is usually in a steady-state fluctuation only during periods when there are large C inputs (Smith et al., 1986; Smith and Paul, 1990).

Carbon mineralization can be expressed on a mechanistic basis by relating product formation to growth and maintenance using the steady-state model depicted in Figure 3 (Smith et al., 1986). Equation 3 describes product formation at steady state

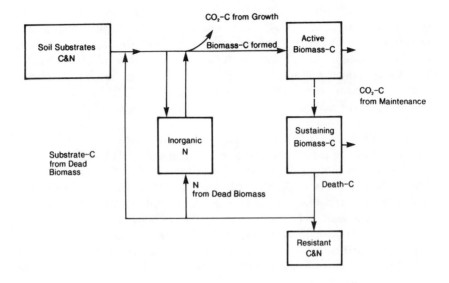

Figure 3. Soil microbial biomass C and N cycle at steady-state. (From Smith et al., 1986.)

$$\frac{dp}{dt}T = Y_{(p/x)}\frac{dx}{dt}a + ax_T \qquad (3)$$

where dP_T is the CO_2-C production over time, the second term is the CO_2-C from growth of the active biomass fraction (x_a) multiplied by the yield of product per unit of biomass, and the last term is the CO_2-C from maintenance activities where a is the maintenance rate and x_T the total population. The mineralization of N can also be expressed on a mechanistic basis as

$$\frac{dN}{dt} = (2.4\ ax_T\ /\ C_{ns}) + 0.09\ ax_T - (1.6\ ax_T\ /\ C_{nfb}) \qquad (4)$$

where the change in N is based on microbial metabolism (ax_T), the C/N ratio of the soil (C_{ns}) and the C/N ratio of the growing SMB (C_{nfb}). The constants were produced from other known variables (see Smith et al., 1986). Using equation 4 to calculate N mineralized during the steady-state period of a 60 d incubation showed good agreement to measured values (Table 3). Using a C/N ratio of forming biomass of 7.5 showed better agreement than using a ratio of 8.0. To

Table 3. Calculated and experimental values for N mineralized during the steady state period of a 60-d incubation

| | Nitrogen mineralized mg N kg^{-1} soil | | | C/N ratio | |
	Actual	C/N 7.5[a]	C/N 8.0[a]	Biomass formed[b]	Soil
Palouse	5.2	3.5	4.7	8.3	14.5
Warden	6.5	5.8	7.0	7.8	13.0
Walla Walla	5.1	4.7	5.7	7.7	12.9
Ritzville	5.0	5.2	6.1	7.4	11.5
Shano	4.5	5.7	6.5	6.8	10.7

[a] Calculated using a C/N ratio of forming biomass of 7.5 or 8.0.
[b] Calculated using actual N mineralized.
From Smith et al., 1986.

predict the measured N mineralized for each soil the C/N ratio of the biomass ranged from 6.8 to 8.3, not unreasonable values for SMB in soils of varying C/N ratios.

The mechanistic analysis of N mineralization, based on C mineralization for energy and cell growth, assumes the SMB value for the total population is accurate and that the SMB yield (mg SMB/mg C substrate) is known. Figure 4 shows an analysis of N mineralized if the SMB value is varied from 0.5 to 1.5 of the actual value and the yield is varied from 0.3 to 0.60. It is quite evident that the biomass value will affect the calculated N mineralized, in addition the yield value has a much more dramatic effect than the accuracy of the biomass value. Thus in mechanistic models of N transformations the measurement of the microbial biomass is critical for accurate results.

In contrast to mechanistic models process type models of N transformations such as mineralization are usually developed using end product data from field studies or laboratory incubations under controlled conditions. Several models describing N mineralization exist with most of them based on 1st order kinetics. Whether using a chemical kinetic model or an enzyme kinetic model such as Michaelis-Menten the reaction order is actually 2nd order due to the SMB being a catalyst. However, under ideal conditions such as optimum temperature and moisture the SMB approaches a maximum and the kinetic equations revert to 1st order (Smith and Paul, 1990). In a survey of 14 N simulation models de Willigen (1991) found that the number of pools for the mineralization/ immobilization reactions ranged from 1 to 7 with most using 1 to 2 pools. Microbial mediated processes were described with 1st order equations (11 models) or 0 order kinetics (3 models). Only five simulation models considered an SMB pool explicitly in the N transformation equations. This is most likely

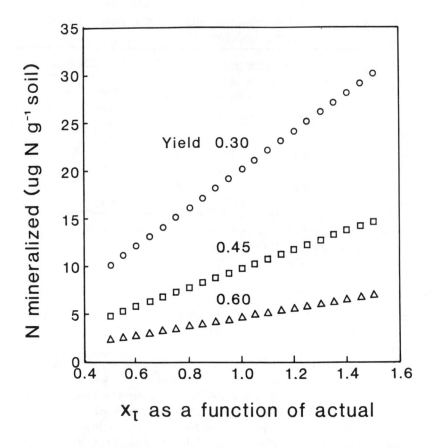

Figure 4. The effect of the SMB value on the mechanistic calculation of N mineralization at three different yield values. (From Smith, 1989)

due to the lack of temporal data for the SMB pool and the more expedient method of incorporating the SMB into a psedo-1st order equation. A conclusion from comparison of the models was that more complex models did not give any better N mineralization results than did simpler models. For further recent advances consult the N cycling and modelling compilations by Groot et al., (1991) and Jenkinson and Smith, (1988).

For modeling N mineralization at least two pools of organic substrates are used for the decomposition processes. Figure 5 is a simplified model of SOM transfers within a soil plant system. If only soil pools are considered an active and resistant (humified) pool of SOM is available for mineralization processes. If plant residue is also considered then 3 more SOM pools, soluble, cellulose and lignin plant material are decomposing simultaneously. In many models of N cycling N is tied to C flux and N mineralization is based on the C/N ratio of

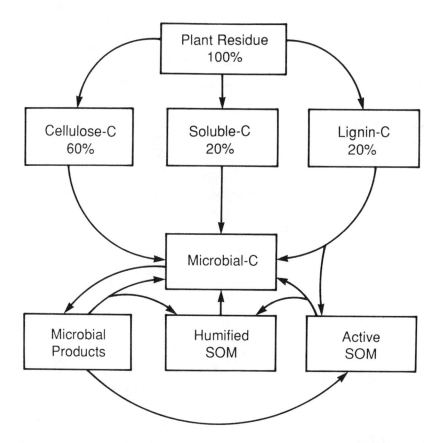

Figure 5. A simple compartmentalized model of microbial mediated mineralization and decomposition.

the pool. Since the activity of the SMB is for the consumption of organic substrates it is not surprising that C and N mineralization should be related to SMB and to the C/N ratio of specific pools. In an experiment to determine geostatistical spatial relationships of microbial processes in a shrub-steppe ecosystem we measured net N mineralization and CO_2 respiration on 205 samples during a 30 day laboratory incubation. On separate field moist samples we measured SMB by the SIR method. Figure 6 shows the linear correlation between the processes and the SMB values. All of the correlations are significant however forcing the regression through zero doubled the r^2 value. On a pair wise basis there was also a significant relationship between the CO_2 evolved per unit of N mineralized. Even though the sampling area was small (200 m²) and the processes and SMB range an order of magnitude the relationships can be used to estimate resource islands and nutrient cycling (Halvorson et al., 1992).

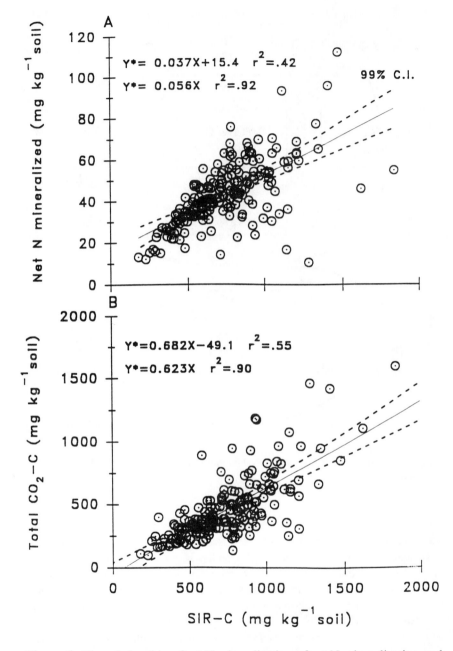

Figure 6. The relationship of net N mineralization of net N mineralization and respiration to soil microbial biomass; 205 soil samples were taken from shrub-steppe ecosystem.

These relationships strengthen the assumption that N mineralization is tied to C mineralization and can be modelled as a coupled reaction.

The coupling of C and N mineralization is reasonable, however if the C pool is not readily metabolized by the SMB then N mineralization will be slow regardless of the C/N ratio of the substrate. Thus the concept of substratequality is introduced as the controlling factor on microbial growth and N mineralization. Figure 7 shows CO_2-C, SMB-C and N mineralized during a 100 day incubation of soil from a 7 year old clear-cut, a 3 year old clear-cut and an annual grassland in California. There is a greater rate of CO_2 production in the 7 year old clear-cut as compared to the other two systems as calculated by 1st order kinetics. What is evident from this data is that there is less SMB-C in the 7 year clear-cut but greater CO_2 production suggesting a substrate quality difference. Comparing metabolic quotients (CO_2-C/SMB-C) showed similar values for the 3 year old clear-cut and the grassland soil but a 48 to 55% increase in the metabolic quotient for the 7 year clear-cut site. The N mineralized per unit of SMB was similar in each soil, however the amount of CO_2 produced per unit of N mineralized was greatest in the 7 year old clear-cut. These metabolic function ratios suggest that the C substrates in the 7 year old clear-cut are of lower quality and that the SMB must metabolize at a greater rate to satisfy their energy and cell building needs. The total N mineralized during the incubation was greatest in the grassland soil, followed by the 3 and 7 year old clear-cuts.

In addition to the quantity and quality of substrates the SMB size and activity is affected by abiotic factors such as temperature and moisture and soil properties such as bulk density, pH, and clay content. For example, Figure 8 from van Veen et al., (1985) shows the effect of texture on the protection of inorganic N incorporated into the organic fraction. In this experiment the clay loam soil had a lower rate of N mineralization than did the sandy loam soil. This clay or texture protection characteristic will also affect SMB activity and turnover. Ladd and Foster (1988) showed a substrate and a soil texture effect on the turnover of C and N through the SMB

It has long been realized that disturbance through cultivation will decrease SOM levels (Jenny, 1941) but tillage also has an effect on the SMB levels (Granatstein et al., 1987; Follett and Schimel, 1989). Bolton et al.(1990) showed that disturbance from farming, halted in the 1940's, still had an effect on the SMB and nitrogen dynamics of a natural ecosystem.

It is evident that a number of factors affect the N mineralization process and the factors controlling N mineralization in a forest ecosystem may be different than the controlling factors in an agroecosystem. It is these differences that are important in ecosystem analysis of the N contribution to plant growth. Using *in situ* or laboratory analysis of metabolic process quotients will enable us to evaluate the health and productivity of the system.

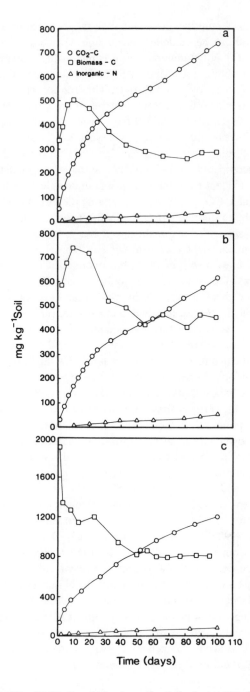

Figure 7. Total CO_2, SMB-C and N mineralized during a 100 d incubation from a) soil from a 7 year old clear-cut, b) a 3 year old clear-cut, and 3) an annual grassland soil.

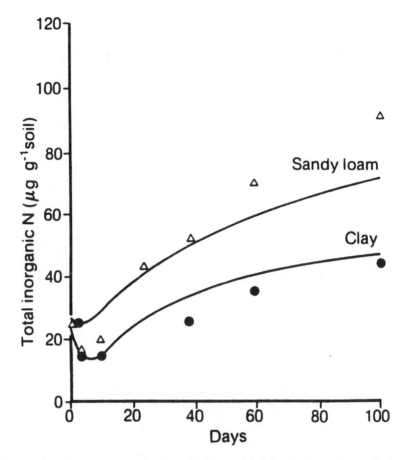

Figure 8. Nitrogen mineralization during a 100 d incubation of a sandy loam and clay soil. (From van Veen et al., 1985.)

V. Microbial Cycling and Plant Uptake of N

In most developed ecosystems plant productivity is dependent upon the mineralization of macronutrients from SOM. Microbially mediated process supplies N,P and S for plant growth and is considered efficient if plant uptake is equal to the net mineralization of nutrients from SOM. However the elemental cycles of N,P and S are not cyclic and losses and immobilizing reactions due occur. Figure 9 depicts the internal soil mineralization cycle showing the fates of N,P and S. The mineralization of plant residues and SOM produces inorganic ions of N,P and S all of which can be utilized by plants, sorbed or precipitated and leached from the system. In most soils NH_4 is readily oxidized to NO_3 which can be reduced to a gas and lost. These nutrient cycles are not as tight in agricultural ecosystems as in natural ecosystems mainly due to added inputs and

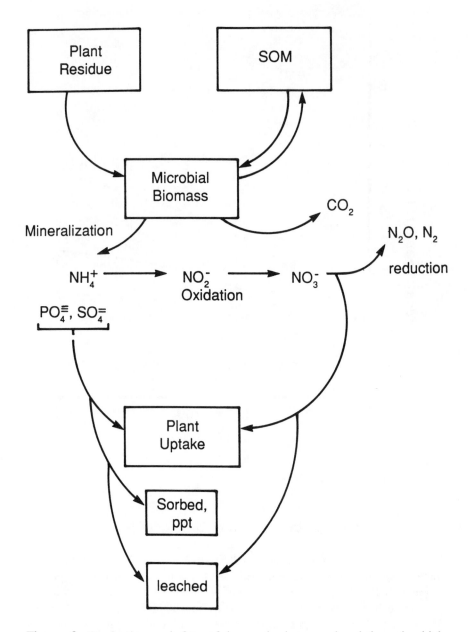

Figure 9. Production and fate of inorganic ions produced by microbial mineralization of plant residue and SOM.

disturbance of the soil. It is our objective in agriculture to tighten these cycles through management in order to reduce the risk of pollution.

To describe the relationship between SMB and plant growth and nutrition I will introduce the hypothesis that the N available for plant uptake comes directly from the SMB and is dependent upon the turnover of the SMB pool (Smith and Paul, 1990). Considering Figure 9 the rate of SOM and plant residue decomposition is slow compared to the turnover of the SMB thus the N released directly into the inorganic N pool is mainly a result of SMB turnover. If this hypothesis is correct then plant productivity will be governed by the size and activity of the SMB.

As previously discussed the size and activity of the SMB is directly related to the quantity and quality of the C input from plants. Thus we have a cyclic cause and effect symbiotic relationship between SMB and plants. Living plants release substantial amounts of photoassimilated C through their roots into the rhizosphere which is readily used by the SMB for growth until the supply is exhausted (Lynch and Whipps, 1990). With regard to a symbiotic type of relationship is it wise for a plant to release C and thus indirectly immobilize nutrients into the rhizosphere SMB? Figure 10 shows two scenarios for C flux from the root to the rhizosphere. If organisms only use the exudates for growth then they are in direct competition with plants for nutrients, a poor strategy for the plant. If on the other hand plant exudates stimulate population activity as well as growth then nutrients will cycle more rapidly increasing the potential uptake by the plant. A brisk debate is currently ongoing as to whether roots have an effect on the mineralization of SOM (Bottner et al., 1988; Sallih and Bottner, 1988; Cheng and Coleman, 1990) which if so should also affect the N mineralization (Merckx et al., 1987; Robinson et al., 1989; Breland and Bakken, 1991).

The competition vs cooperation hypothesis has no means been settled, however examine Figure 11 showing SMB and inorganic N at distances from fine roots from a ^{14}C pine tree labeling experiment. It is quite evident that SMB and N are greater near the root as is the radioactivity (Sp. Act.) of the SMB indicating a more cooperative scenario than competition (also see Norton et. al., 1990). For an excellent review of C flow and the current hypotheses consult Dormaar, (1988,1990).

In the winter wheat growing region of the Pacific Northwest, USA a grain production of 6.7 tonne translates to 16 tonnes ha[-1] of total dry matter plant production. The amount of N, P and S contained in this plant matter is 302, 36 and 32 kg ha[-1] of N,P and S respectively. The average SMB concentration of these elements is 180, 17, and 9 kg ha[-1] of N, P, and S respectively. Thus the SMB contains 60% of the N, 47% of the P and 28% of the S needed to produce this crop. Thus a rapid turnover of the SMB could indeed supply 100% of the crop nutritional requirements.

The above calculation is based on average values and not temporal fluctuations. Figure 12 shows the temporal flux of SMB and nitrate in comparison to winter wheat N uptake. SMB and NO_3 have peaks in the early spring and late

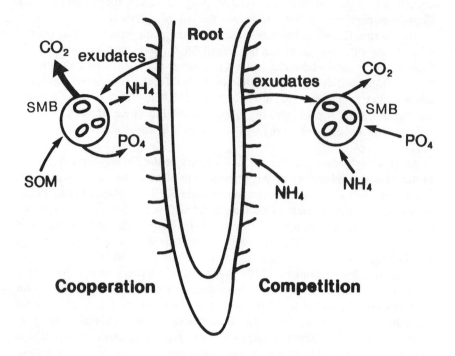

Figure 10. Schematic illustration of the hypothesis of cooperation or competition between the SMB and plants.

fall. When plant uptake of N is rapidly increasing in the spring both NO_3 and SMB are declining suggesting that the pool for plant available N is the SMB. However, Patra et al.(1990) showed little change in the SMB pool over 1.5 years in a wheat and grassland soil.

Figure 13 shows the results of a field study of N cycling in winter wheat where we measured plant, SMB and inorganic N over the growing season at two N levels a) 46 and b) 107 kg-N ha^{-1}. Inorganic N and SMB-N dominated the system in the late fall but continued to decline as plant uptake increased through May. The August values are for grain N and inorganic N showing that after plant maturity inorganic N, mainly NO_3 begins to increase. In a similar field experiment with residue management and N treatments with and without plants we measured SMB-N. Figure 14 shows that in all treatments the planted plots had lower levels of SMB-N than did plots without plants. It appears that as the SMB cycles plants are successfully competing for the released N. In this experiment the C data (not shown) also suggests that the C/N ratio of the SMB in the planted plots was greater than the plots without plants.

In a laboratory experiment using ^{14}C labeled glucose, cellulose and ^{15}N labeled NO_3, Nicolardot (1988) showed a very rapid immobilization of ^{15}N and

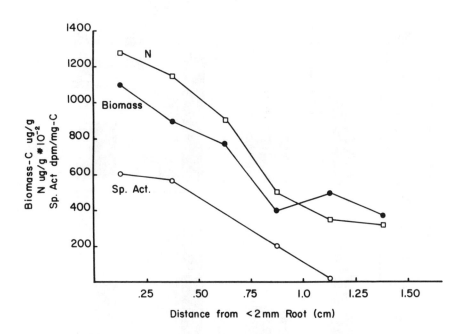

Figure 11. SMB-C, inorganic N and specific activity of the SMB at distances from <2 mm roots of a pine seedling after exposure to $^{14}CO_2$.

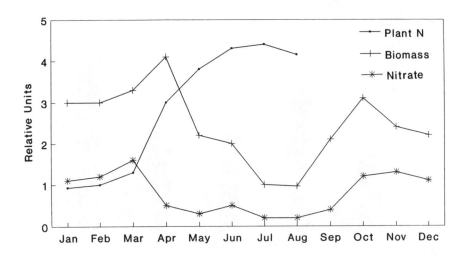

Figure 12. Temporal fluctuation of SMB and NO_3 as related to winter wheat uptake of N in the Pacific Northwest.

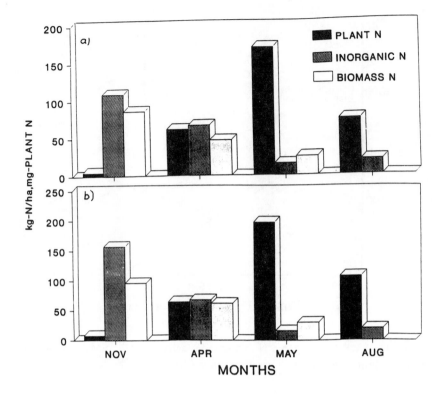

Figure 13. Distribution of N between winter wheat, inorganic N and SMB-N during the growing season: a) 46 kg N ha^{-1} added, b) 107 kg N ha^{-1} N added.

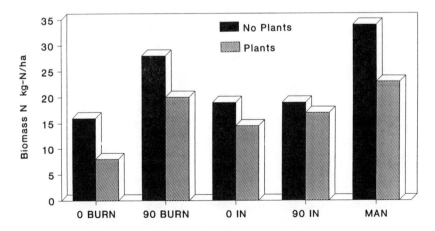

Figure 14. SMB-N in plots with and without plants and treatments of fertilizer 0 or 90, and residue burning or incorporated (N) and manure (MAN). Samples were taken at the height of plant N uptake in April.

mineralization of ^{14}C by the SMB. By day 7 about 50% of the C substrates were mineralized and the N immobilized by day 3 and 7 in the glucose and cellulose treatments respectively. Between 30 and 40% of the newly immobilized N was remineralized during 12 weeks of incubation. The net mineralization for the three soils ranged from 30 to 60 mg kg^{-1} over an 80 day period. Nicolardot concluded that the net N mineralized was too large to come exclusively from the SMB.

Figure 15 shows the SMB-N from that experiment measured over a 250 day period. I have added to the figure the net mineralized N from treated soil (T) and the control soil (C) for the 80 day period described in the paper. I have also calculated the increase in inorganic N between the treated and control soil. I then estimated from the graphs the difference in the N flush from 0 to 80 days assuming both C treatments to be the same at that time. The data in Figure 15 show that the increase in inorganic N is parallel to the decrease in SMB-N. The low value for mineralized N compared to the change in N flush for the Leuven soil maybe due to the higher C/N ratio of that soil compared to the other two.

For the SMB pool to supply plants with sufficient nutrients for growth the rate of nutrient availability becomes critical. The movement of nutrients from the SMB to plants is facilitated by the turnover rate of the SMB (van Veen et al., 1987; Jenkinson and Parry, 1989). Various turnover times for SMB have been published ranging from less than 1 year based on steady state conditions (Smith and Paul, 1990) to greater than 6 years in a cold climate (Paul and Voroney, 1984). Chaussod et al., (1988) with several soil treatments found that the SMB turnover rate averaged 0.41 years in laboratory incubations and 1.62 years in field measurements, citing temperature differences as the cause. The measurement of SMB turnover times is problematic due to the numerous pools of C in soil systems. Typical approaches have included adding soluble substrates such as ^{14}C labeled glucose to soil or measuring ^{14}C in the biomass from the decomposition of labeled plant material. It is recognized that caution should be taken when calculating turnover rates from single compartment models (Jenkinson and Ladd, 1981; Chaussod et al., 1988).

After determining an SMB turnover rate the flux of nutrients moving through the microbial biomass is then calculated by dividing the pool by the SMB turnover rate. For the example given above for winter wheat the SMB turnover rate would have to be between 0.3 and 0.6 to equal the crop uptake of nutrients. Paul and Voroney (1984) showed that the N flux through the SMB was sufficient to supply wheat crops in Canada and Rothamsted and a sugar crop in Brazil. Lethbridge and Davidson (1982) also found that the SMB could act as the sole source of N for wheat and maize plants. Based on steady-state assumptions (i.e. litter input = decomposition) we have estimated the SMB turnover time to range from 0.2 to 0.6 years for several global vegetation types (Smith and Paul, 1990). For each of these systems the flux of nutrients is greater than the plant biomass uptake. Table 4 gives the global comparison of C,N,P,and S pools. In the case of N the SMB turnover rate must be less than 1 year to satisfy plant uptake, however for P and S the rate could be greater than 1 year.

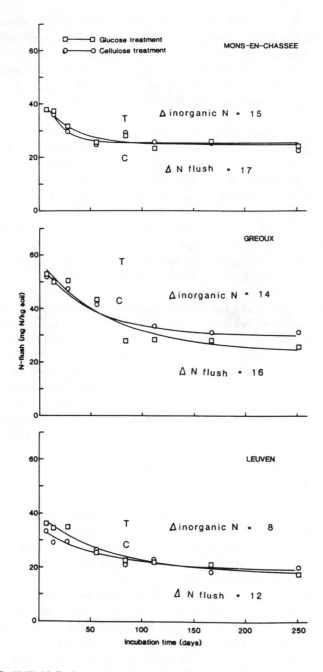

Figure 15. SMB-N flush measured periodically over a 250 d incubation period: inorganic N is the amount of N mineralized over an 80 d period and the delta N flush is the difference between the first N flush and the 80 d N flush. (Modified from Nicolardot, 1988.)

Table 4. Comparison of global pool magnitudes and yearly nutrient uptake of C, N, P, and S by plant biomass, in Gt

	C	N	P	S
Plant biomass	785	13.0	1.8	1.4
Yearly uptake	60	1.2	0.2	0.1
Microbial biomass	6	0.9	0.7	0.2

From Smith and Paul, 1990.

Conditions vary from crop to crop and certainly locations differ in soil type and climate, but it seems that more and more data being collected is substantiating the hypothesis that the N flux from the SMB is the major source of plant N. The data presented above is by no means exhaustive and examples are quite abundant in the literature many of which have not been interpreted.

VI. Conclusions

Nitrogen cycling through the SMB has a major influence on plant productivity and ecosystem development. The distribution of N in the soil system is largely a function of microbial biomass size and activity. The emerging concept of soil quality or health will certainly have as one of the major focuses the SMB and it's functioning in relationship to N transformations.

Even though process models describing mineralization/immobilization are less complex than mechanistic based models there is still a need to measure SMB along with specific processes. The major problems surrounding the CFIM biomass method could be eliminated by simply subtracting 20% of the total CO_2 evolved from the control sample. If one was to only use a small number of soils in their research program it would be advantageous to use the ^{14}C leached straw method to determine a specific correction factor for those soils. In addition I would advocate that SMB-N be reported as the N flush in scientific publication in order for comparison to other data.

There seems to be evidence that microbial release of N coincides with the plant's need for N at least in some agricultural areas. The question remains, can we manage the SMB and it's release of N to make agriculture more efficient and ecologically sound. The ideal scenario would be to have the SMB absorb and hold significant amounts of N releasing it at the time of peak plant uptake. This would also involve increasing the SMB at particular times of the year. Perhaps evaluating natural ecosystems with tighter N cycles would provide a model for managing the SMB and it's activity.

There are several future research needs and questions concerning SMB and the N cycle that warrant listing.

1. Develop a chemical measurement of active soil N.
2. Is it possible to increase the active soil N fraction?
3. Can we manage the mineralization/immobilization cycle?
4. Elucidate the relationship between SMB and the active N pool.
5. Define the relationship between organism diversity and the N cycle.
6. How does global change affect SMB N cycling?

References

Anderson, J.P.E. and K.H. Domsch. 1978. A physiological method for the quantitative measurement of microbial biomass in soils. *Soil Biol. Biochem.* 10:215-221.

Anderson, T. and K.H. Domsch. 1989. Ratios of microbial biomass carbon to total organic carbon in arable soils. *Soil Biol. Biochem.* 21:471-479.

Bolton, H. Jr., J.L. Smith, and R.E. Wildung. 1990. Nitrogen mineralization potentials of shrub-steppe soils with different disturbance histories. *Soil Sci. Soc. Am. J.* 54:887-891.

Bottner, P., Z. Sallih, and G. Billes. 1988. Root activity and carbon metabolism in soils. *Biol. Fertil. Soils* 7:71-78.

Breland, T.A. and L.R. Bakken. 1991. Microbial growth and nitrogen immobilization in the root zone of barley (Hordeum vulgare L.), Italian ryegrass (Lolium multiflorum Lam.), and white clover (Trifolium repens L.). *Biol. Fertil. Soils* 12:154-160.

Brookes, P.C., A. Landman, G. Pruden, and D.S. Jenkinson. 1985. Chloro form fumigation and the release of soil nitrogen: a rapid direct extraction method to measure microbial biomass nitrogen in soil. *Soil Biol. Biochem.* 17:837-842.

Chaussod, R., S. Houot, G. Guiraud, and J.M. Hetier. 1988. Size and turnover of the microbial biomass in agricultural soils: Laboratory and field measure-ments. p. 312-326. In: D.S. Jenkinson and K.A. Smith (eds.), Nitrogen efficiency in agricultural soils. New York: Elsevier Science Publishing Co.

Cheng, W. and D.C. Coleman. 1990. Effect of living roots on soil organic matter decomposition. *Soil Biol. Biochem.* 22:781-787.

de Willigen, P. 1991. Nitrogen turnover in the soil-crop system; comparison of fourteen simulation models. In: J.J.R. Groot, P. de Willigen, and E.L.J. Verberne (eds.). Nitrogen turnover in the soil-crop system. Kluwer Academic Publishers, The Netherlands.

Dormaar, J.F. 1988. Effect of plant roots on chemical and biochemical properties of surrounding discrete soil zones. *Can. J. Soil Sci.* 68:223-242.

Dormaar, J.F. 1990. Effect of active roots on the decomposition of soil organic materials. *Biol. Fertil. Soils* 10:121-126.

Follett, R.F. and D.S. Schimel. 1989. Effect of tillage practices on microbial biomass dynamics. *Soil Sci. Soc. Am. J.* 53:1091-1096.

Geyer, D. J., C. K. Keller, J. L. Smith, and D. L. Johnstone. 1992. Subsurface fate of nitrate as a function of depth and landscape position in Missouri Flat Creek watershed. *J. Contam. Hydrol.* (In press).

Granatstein, D.M., D.F. Bezdicek, V.L. Cochran, L.F. Elliott, and J. Hammel. 1987. Long-term tillage and rotation effects on soil microbial biomass, carbon and nitrogen. *Biol. Fertil. Soils* 5:265-270.

Gregorich, E.G., G. Wen, R.P. Voroney, and R.G. Kachanoski. 1990. Calibration of a rapid direct chloroform extraction method for measuring soil microbial biomass C. *Soil Biol. Biochem.* 22:1009-1011.

Groot, J.R., P. de Willigen, and E.L.J. Verberne (eds.). 1991. Nitrogen turnover in the soil-crop system. Norwell, MA: Kluwer Academic Publishers.

Halvorson, J.J., E.H. Franz, J.L. Smith, and R. A. Black. 1991. Nitrogenase activity, dinitrogen fixation and nitrogen input by lupines at Mt. St. Helens. *Ecology.* 73:87-98.

Halvorson, J.J., H. Bolton Jr., J.L. Smith, and R.E. Rossi. 1992. Measuring resource islands using geostatistics. *Ecology* (In review).

Jenkinson, D.S. 1988. Determination of microbial biomass carbon and nitrogen in soil. p. 368-386. In: J.R. Wilson (ed.), Advances in nitrogen cycling in agricultural ecosystems. Wallingford, UK.: C.A.B International.

Jenkinson, D.S. and J.N. Ladd. 1981. Microbial biomass in soil: measurement and turnover. In: E.A. Paul and J.N. Ladd (eds.), *Soil Biochemistry* vol. 5. pp.415-471. Marcel Dekker, New York.

Jenkinson, D.S., and D.W. Powlson. 1976. The effects of biocidal treatments on metabolism in soil-V. A method for measuring soil biomass. *Soil Biol. Biochem.* 8:209-213.

Jenkinson. D.S. and K.A. Smith. 1988. *Nitrogen efficiency in agricultural soils.* Elsevier Science Publishing Co., New York, NY.

Jenkinson, D.S. and L.C. Parry. 1989. The nitrogen cycle in the broadbalk wheat experiment: A model for the turnover of nitrogen through the soil microbial biomass. *Soil Biol. Biochem.* 21:535-541.

Jenny, H. 1941. Factors of soil formation, a system of quantitative pedology. p.281. , New York: McGraw-Hill.

Ladd, J.N. and R.C. Foster. 1988. Role of soil microflora in nitrogen turnover. p. 113-133. In: J.R. Wilson (ed.), Advances in nitrogen cycling in agricultural cosystems. Wallingford, UK.: C.A.B International.

Lethbridge, G., and M.S. Davidson. 1982. Microbial biomass as a source of nitrogen for cereals. *Soil Biol. Biochem.* 15:375-376.

Lynch, J.M. and J.M. Whipps. 1990. Substrate flow in the rhizosphere. *Plant and Soil* 129:1-10.

Merckx, R., A. Dijkstra, A. Den Hartog, and J.A. van Veen. 1987. Production of root-derived material and associated microbial growth in soil at different nutrient levels. *Biol. Fertil. Soils* 5:126-132.

Nicolardot, B. 1988. Behavior of newly immobilized nitrogen in three agricultural soils after addition of organic carbon substrates. p. 312-326. In: D.S. Jenkinson and K.A. Smith (eds.), Nitrogen efficiency in agricultural soils. New York: Elsevier Science Publishing Co.

Norton, J. M., J. L. Smith, and M. K. Firestone. 1990. Carbon flow in the rhizosphere of ponderosa pine seedlings. *Soil Biol. Biochem.* 22:449-445.

Ocio, J.A. and P.C. Brookes. 1990. Soil microbial biomass measurements in sieved and unsieved soil. *Soil Biol. Biochem.* 22:999-1000.

Parkinson, D. and E.A. Paul. 1982. Microbial biomass. p. 821-830. In: A. L. Page, R. H. Miller, and D. R. Keeney (eds.), Methods of soil analysis. Part 2. Chemical and microbiological properties. Madison, WI.: Agronomy Monograph no. 9, ASA-SSSA.

Patra, D.D., P.C. Brookes, K. Coleman, and D.S. Jenkinson. 1990. Seasonal changes of soil microbial biomass in an arable and grassland soil which have been under uniform management for many years. *Soil Biol. Biochem.* 22:739-742.

Paul, E.A. and R.P. Voroney. 1984. Field interpretation of microbial biomass activity measurements. p. 509-515. In: M.J. Klug and C.A. Reddy (eds.), Current perspectives in microbial ecology. Washington, D.C.: American Society for Microbiology.

Ritz, K. and R.E. Wheatley. 1989. Effects of water amendment on basal and substrate-induced respiration rates of mineral soils. *Biol. Fertil. Soils* 8:242-246.

Robinson, D., B. Griffiths, K. Ritz, and R. Wheatley. 1989. Root-induced nitrogen mineralisation: A theoretical analysis. *Plant and Soil* 117:185-193.

Ross, D.J. 1989. Estimation of soil microbial C by a fumigation-extraction procedure: Influence of soil moisture content. *Soil Biol. Biochem.* 21:767-772.

Ross, D.J., K.R. Tate, A. Cairns, and K.R. Meyrick. 1980. Influence of storage on soil microbial biomass estimated by three biochemical procedures. *Soil Biol. Biochem.* 12:369-374.

Rosswall, T. 1976. The internal N cycle between microorganisms, vegetation and soil. p. 157-168. In: B.H. Svensson and R. Soderlund (eds.), Nitrogen,phosphorus and sulphur-global cycles. *Ecol Bull.* Vol 22. Stockholm.

Sallih, Z. and P. Bottner. 1988. Effect of wheat (Triticum aestivum) roots on mineralization rates of soil organic matter. *Biol. Fertil. Soils* 7:67-70.

Schnurer, J., M. Clarholm, and T. Rosswall. 1985. Microbial biomass and activity in an agricultural soil with different organic matter contents. *Soil Biol. Biochem.* 17:611-618.

Smith, J.L.. 1989. Sensitivity analysis of critical parameters in maintenance energy models. *Biol. Fertil. Soils* 8:7-12.

Smith, J.L. and L.F. Elliot. 1990. Tillage and residue management effects on soil organic matter dynamics in semi-arid regions. p. 69-88. In: B. A. Stewart (ed.) *Advances in Soil Science,* vol 13. Springer Verlag, New York, N.Y.

Smith, J.L., B.L. McNeal, H.H. Cheng, and G.S. Campbell. 1986. Calculation of microbial maintenance rates and net nitrogen mineralization in soil at steady state. *Soil Sci. Soc. Am. J.* 50:332-338.

Smith, J.L. and E.A. Paul. 1986. The role of soil type and vegetation on microbial biomass and activity. p. 460-466. In: F. Magusar and M. Gantar (eds.), Perspectives in microbial ecology. Slovene Society for Microbiology, Ljubljana.

Smith, J.L. and E.A. Paul. 1990. The significance of soil microbial biomass estimations. p. 357-396. In: J.M. Bollag and G. Stotzky (eds.), *Soil Biochemistry.* Marcel Dekker, New York, N.Y.

Smith, J.L., J.M. Lynch, D.F. Bezdicek, and R.I. Papendick. 1992. Soil organic matter dynamics and crop residue management. In: B. Metting (ed.), *Soil microbial ecology.* Marcel Dekker, New York, N.Y. (in press.)

Soderlund, R. and B.H. Svensson. 1976. The global nitrogen cycle. P. 23-73. In: B.H. Svensson and R. Soderlund (eds.), Nitrogen, phosphorus and sulphur-global cycles. *Ecol. Bull.* Vol. 22. Stockholm, Sweden.

Sparling, G.P. and A.W. West. 1988. A direct extraction method to estimate soil microbial C: Calibration *in situ* using microbial respiration and ^{14}C labelled cells. *Soil Biol. Biochem.* 20:337-343.

Stevenson, F.J. 1982. Origin and distribution of N in soil. p. 1-42. In: F.J. Stevenson (ed.), Nitrogen in agricultural soils. American Society of Agronomy, Madison, WI.

Tate, K.R., D.J. Ross, and C.W. Feltham. 1988. A direct extraction method to estimate soil microbial C: Effects of experimental variables and some different calibration procedures. *Soil Biol. Biochem.* 20:329-335.

van Veen, J.A., J.N. Ladd, and M. Amato. 1985. Turnover of carbon and nitrogen through the microbial biomass in a sandy loam and a clay soil incubated with [^{14}C(U)] glucose and [^{15}N] (NH$_4$)$_2$SO$_4$ under different regimes. *Soil Biol. Biochem.* 17:747-756.

van Veen, J.A., J.N. Ladd, J.K. Martin, and M. Amato. 1987. Turnover of carbon, nitrogen, and phosphorus through the microbial biomass in soils incubated with ^{14}C-, ^{15}N- and ^{32}P-labelled bacterial cells. *Soil Biol. Biochem.* 19:559-565.

Voroney, R.P. and E.A. Paul. 1984. Determination of k_C and k_N *in situ* for calibration of the chloroform fumigation-incubation method. *Soil Biol. Biochem.* 16:9-14.

Whittaker, R.H. and G.E. Likens. 1973. Carbon in the biota. p. 281-302. In: G.M. Woodwell and E.V. Pecan (eds.), Carbon and the biosphere. Proceedings of the 24th Brookhaven Symposium in Biology, CONF-720510. National Technical Information Service, Springfield, VA.

Zinke, P.J., A.G. Strangenberger, W.M. Post, W.R. Emanuel, and J.S. Olson. 1984. Worldwide organic soil C and N data. In: Environmental science division publication N. 2212. Oak Ridge National Laboratory, Oak Ridge, TN.

Pesticide Degradation by Soil Microorganisms: Environmental, Ecological, and Management Effects

Thomas B. Moorman

I. Introduction

Pesticides are used principally in agricultural environments to control weeds, insects, parasitic fungi, and nematodes. Enormous amounts of pesticides are used globally; 341 million kg in agriculture in the USA alone (Pimental and Levitan, 1986). Much of these products are either applied directly to soil, or indirectly as in the case of postemergence herbicides. The ability of the soil to detoxify, bind or retard the movement of these chemicals into the air or water helps reduce environmental contamination. The detoxification and ultimate fate

This chapter was prepared by a U.S. government employee as part of his official duties and legally cannot be copyrighted.

of these pesticides and their metabolites is principally mediated by microorganisms, although abiotic processes are important for some compounds. These microbial processes are important in maintaining the capacity of the soil to assimilate these compounds, thus maintaining the quality of the soil resource.

The need to reduce pesticide impacts on soil and offsite habitats has driven the continued scientific interest in the biodegradation of pesticides and related compounds. Aspects of pesticide biodegradation range from the illucidation of individual steps in the degradation pathways to the complex mathematical simulation models of the biodegradation process that predict pesticide movement and fate. Previous reviews have focused on aspects of the biochemistry and genetics of pesticide degradation by isolated microoorganisms (Munnecke et al., 1982; MacRae, 1989). Alternatively, other reviews have examined the rates of pesticide degradation in soil or water systems with little specific attention to the microbiological component (Hance, 1984; Yaron et al., 1985). The objectives of this discussion are to examine the physiological and ecological factors that are most important in controlling the biodegradation process in agricultural and natural environments, with particular emphasis on the application of this knowledge to the problems of soil and water quality. An understanding of how soil microorganisms function within the soil environment may offer new ways to manage the degradation process in beneficial ways.

II. Biochemistry and Pathways of Degradation

A. Transformation Pathways

The biochemistry of the biodegradation process affects the rates of degradation and the types of metabolites that are produced. Aromatic derivatives of benzoate, phenol, aniline, triazine, pyrolles, pyrimidines and quinoline can all be found singly or in combination as the basic structure of different pesticides (Figure 1). Chemical names for pesticides referred to in this chapter are listed in Appendix A. These basic aromatic ring structures are modified by various substituents, such as ring-substituted halogens, amines, trifluoromethyl groups and organic groups which often have substantial effects on their pesticidal efficacy. Increasing the numbers and types of ring substituents and organic groups often increases the recalcitrance of the molecule in soils and aquatic sediments. For instance, nitrobenzene required 64 days for decomposition by a mixed soil population compared to 4 days for aniline and 1 day for benzoate and phenol (Alexander and Lustigman, 1966). The effects of structure are reflected ultimately as differences in persistence in the environment. Half-lives of 2,4-D were less than 7 days in two soils, while the more chlorinated 2,4,5-T had half-lives of 13 days (Smith, 1978).

Pesticide degradation is usually initiated by enzymatic modification of the non-aromatic portion of the molecule. Reactions include hydroxylation, *N*-dealkylation, decarboxylation, dechlorination, beta-oxidation, ester-hydrolysis

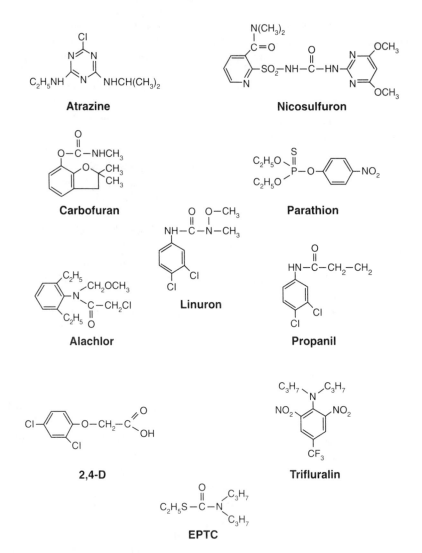

Figure 1. Pesticide chemical structures representative of major chemical classes: acetanilides (propanil and alachlor), carbamothioates (EPTC), dinitroanilines (trifluralin), methylcarbamates (carbofuran), organophosphates (parathion), phenoxyalkanoic acids (2,4-D), substituted-ureas (linuron), sulfonylureas (nicosulfuron), and triazines (atrazine).

and sulfoxidation (Bollag and Liu, 1990). For instance, substituted-urea and many carbamate herbicides are converted in single or multiple steps to chlorinated aniline metabolites (Kaufman, 1967; Lanzilotta and Pramer, 1970; Kaufman and Blake, 1973). This incomplete degradation of a pesticide by microorganism has been termed biotransformation. In some cases, such as the degradation of propanil to propanoic acid and 3,4-dichloroaniline, the propanoic acid group is sufficient to support limited microbial growth (Lanzilotta and Pramer, 1970).

In soils, pesticide biotransformations result in accumulations of some metabolites for short periods after application. For instance, substituted-urea herbicides are metabolized by sequential demethylation and hydrolysis of the urea linkage to form a substituted aniline. Mueller et al. (1992) found that surface soil concentrations of desmethyl-fluometuron increased to approximately 250 ng g^{-1} following application of 1500 ng g^{-1} of fluometuron, but after 100 days desmethyl-fluometuron was not detected. Only trace amounts of 3-(trifluoromethyl)phenylurea were measured in the same study. Concentrations of 3,4-dichloroaniline one year after the last application of linuron or diuron were less than 25 ng g^{-1}, even after several years of annual herbicide application (Belasco and Pease, 1969; Khan et al., 1976).

Other chemicals are biotransformed with no apparent advantage to the microorganism in terms of carbon incorporation or energy production. For example, monochlorinated anilines did not support growth of *Rhodococcus* strains (Janke et al., 1989), although these strains could grow on aniline. The amounts of these substrates converted to chlorocatechols were dependent upon the availability of a secondary substrate (glucose) as an energy source. Dead-end products *cis*-4-carboxymethylene-but-2-en-4-olide and 2-chloro-*cis*,*cis*-muconic acid were produced in cultures. These types of reactions have been referred to as cometabolism (Hovarth, 1972). It is important to note that the absence of growth during pesticide degradation is not necessarily indicative of cometabolism; growth may not be possible due to other nutrient requirements or the low carbon content in trace amounts of pesticide residues may not support growth.

The complete degradation of an organic chemical to CO_2 (mineralization) and other mineral constituents is the result of microbial activity, either as individuals or as consortia. Mineralization of pesticide ring structures (assessed by $^{14}CO_2$ production) indicates the metabolism of the compound or its metabolites with a release of energy and incorporation of pesticide-derived carbon into cellular biomass. However, the evolution of $^{14}CO_2$ from ^{14}C-labeled pesticide is not necessarily evidence of mineralization conducted by individual microorganisms. Complete mineralization often requires multiple enzymatic steps such as the removal of side-chains followed by ring cleavage and metabolism. Microorganisms have been isolated which are able to perform the multiple steps required for complete degradation of many pesticides. The ability to degrade a specific pesticide does not seem to be limited to a particular taxonomic group of microorganisms. Several genera of bacteria have been isolated from soil and aquatic systems which mineralize 2,4-D (see pathway in Figure 2). In soils treated with 2,4-D, the metabolite 2,4-dichlorophenol is produced and subsequently mineralized by other species. This occurs concurrently with 2,4-D mineralization by individual species.

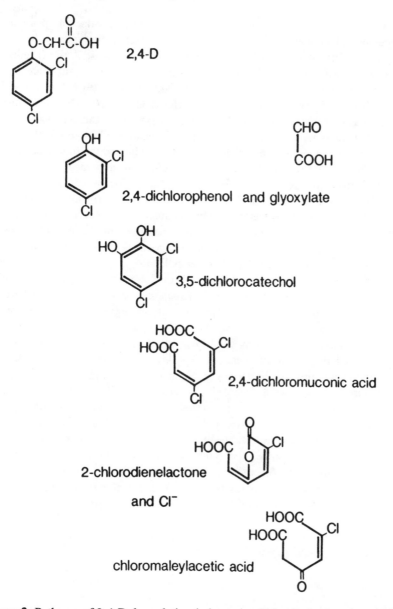

Figure 2. Pathway of 2,4-D degradation in bacteria. Chloromaleylacetic acid is further metabolized to succinate and eventually to biomass and CO_2. In soil, small amounts of 2,4-dinitrophenol accumulate.

The aromatic portion of pesticide molecules, such as substituted benzene rings, phenols, and anilines, are degraded by the ortho- and meta-cleavage pathways in bacteria. The pathways are similar in that oxygenase enzymes convert these compounds to a common intermediate, catechol, prior to ring opening. The opening of the catechol ring by catechol dioxygenase is followed by metabolism that yields biomass and CO_2. The herbicide 2,4-D (Figure 2) is metabolized in this way. In the case of aromatics with nitro substituents, such as the pesticide metabolite dinitrophenol, NO_2^- is produced before cleavage of the phenol ring (Zeyer et al., 1986; Bruhn et al., 1987). The metabolism of chlorinated-aromatic rings is blocked if enzymes in the pathway, such as catechol dioxygenase cannot act on the halogenated substrates. If this occurs, chlorinated metabolites can accumulate in the media and the ring structure is not mineralized. A *Pseudomonas fluorescens* culture accumulated 2,4,5-trichloro-phenol from the partial degradation of 2,4,5-T (Rosenberg and Alexander, 1980). Metabolic blocks are also observed with ring substituents other than halogens. The trifluoromethyl group common to several pesticides such as fluometuron and trifluralin, also prevents aromatic ring metabolism as seen in studies with 3-(trifluoromethyl)benzoate (Engesser et al., 1988). These examples represent additional examples of cometabolism.

Microorganisms can also utilize N, P or S contained in pesticides with or without simultaneous C utilization. A *Pseudomonas* sp. was isolated from soil which degraded the pesticide metabolite 4-chloroaniline as both a source of C and N (Zeyer and Kearney, 1982a). The [^{14}C-ring]aniline was mineralized (64%) and incorporated into biomass (14%). In other studies deethylsimazine, a dealkylated metabolite of simazine, was deaminated to produce an additional triazine metabolite with bacterial growth on the N occuring quantitatively (Cook and Hütter, 1984). Bacteria were also isolated from soils with previous exposure to s-triazine herbicides that converted the sulfur-containing ametryn and prometryne to their respective hydroxylated metabolites (Cook and Hütter, 1982). These bacteria were able to utilize methylthio group as an S source. Cook et al. (1978) showed that microbial growth could be obtained in P-free media on a number of compounds, including pesticide metabolites. Later, Moore et al. (1983) showed that a *Pseudomonas* strain was able to cleave the C-P bond of glyphosate and utilize the P as a sole P source for growth.

B. Enzyme Specificity and Regulation

1. Enzyme Specificity

A number of different types of enzymes are involved in pesticide biodegradation. Several enzyme systems, exclusive of the ortho- and meta-cleavage pathways, have been characterized, including esterases, amidases, oxygenases (cytochrome P-450) and glutathion-s-transferase. Results with purified or partially purified enzyme systems reveal that pesticides of similar structures can be degraded by the same enzyme. This has been shown most conclusively with the organophosphates, the phenoxyalkanoic acids and a large group of related compounds including the acetanilides, carbamates, and phenylureas (Munnecke et al., 1982).

Enzyme specificity is controlled in part by the steric configuration of the substrate molecule. Kearney (1967) summarized the effects of ring substitution on activity of an acylamidase isolated from *Pseudomonas striata*, which hydrolyzes the carbamates CIPC and IPC to short chain organic alcohols, CO_2, and the corresponding anilines. As the size of the organic side-chain increased the rate of hydrolysis decreased by as much 84%. Substitution at the meta or para positions on the ring also affected enzymatic hydrolysis. These results were attributed to electrochemical effects on the ring or the carbamate linkage.

Similar results to those obtained with the *P. striata* acylamidase have been obtained in other systems. The enzyme hydrolyzing *N*-methylcarbamates isolated from *Achromobacter* sp. WM111 had K_m values of 15 for carbaryl, 56 carbofuran, and 2800 μM for aldicarb, demonstrating a strong preference for carbaryl and carbofuran over aldicarb (Derbyshire et al., 1987). The V_{max} values for carbaryl and carbofuran were also greater than for aldicarb. This enzyme is similar to another enzyme hydrolyzing *N*-methylcarbamates from a *Pseudomonas* sp. (Mulbry and Eaton, 1991), although these two enzymes differ in their molecular weights. The failure of these enzymes to act against phenylcarbamates, such as CIPC distinguishes them from acylamidases. The occurrence of similar enzymes in different organisms, such as the amidases isolated fom *Bacillus*, *Pseudomonas*, and different *Fusarium* species (Engelhardt et al., 1973; Kearney, 1967; Marty and Vogues, 1987; Reichel et al., 1991) suggest that these enzymes are widely distributed in both soil bacteria and soil fungi. The ability of these enzymes to degrade a number of structurally similar compounds suggests that they may play a role in the development of enhanced rates of biodegradation.

Two other microbial enzyme systems have recently received further attention for pesticide-degrading potential. Cytochrome P-450 monooxygenase is distributed in both eukaryotes and prokaryotes. This enzyme in a *Streptomyces* sp. has recently been shown to hydroxylate short aliphatic side chains of sulfonylurea herbicides (Sariaslani, 1991). Cytochrome P-450 monooxygenase has long been known to participate in the oxidation and dechlorination of aliphatic hydrocarbons, but its role in the biotransformation of aromatic pesticides has not received much attention.

Glutathione S-transferase is another enzyme that is also present in bacteria and fungi. Glutathione S-transferase was apparently responsible for the formation of oxanilic and sulfonic acid derivatives of acetochlor (Feng, 1991). In these reactions S-glutathione conjugates are formed with concurrent dechlorination of acetochlor. The oxanilic acid is formed by hydrolysis of the S-glutathione and oxidation of the remaining carbon. Alternatively, cleavage of the glutathion peptide followed by oxidation of the remaining sulfur to form the sulfonate metabolite. These metabolic products have also been reported as soil metabolites of alachlor (Sharp, 1988) and a sulfur replacement of Cl was detected during metolachlor metabolism (Liu et al., 1989).

The enzymes discussed previously are intracellular, but there is some evidence that esterases and amidases with pesticide-degrading activity exist as extracellular soil enzymes (Burns and Edwards, 1980; Gaynor, 1992; Nakamura et al., 1992). It is necessary to distinguish between intra- and extra-cellular enzymes, which has been accomplished through treatments that inhibit cellular activity. Gaynor (1992) examined the potential role of extracellular esterases in the conversion of diclofop-methyl to diclofop. Esterase activity in soil treated with propylene oxide or sodium azide retained 30 to 70% of the activity in untreated soil. Purified lipase and esterase enzymes also hydrolyzed diclofop-methyl. At this time the role of extracellular enzymes has not been systematically quantified, but the potential role of these enzymes needs to be recognized.

2. Regulation

Little is known about the exact molecular regulatory mechanisms of pesticide-degrading enzymes. The degradation of pesticides by microorganisms in pure cultures is often marked by lag periods, suggesting that enzyme induction is occurring. The regulation of such enzymes is only partially understood in a few cases. An amidase from *Bacillus sphaericus*, which degrades a broad variety of phenylurea pesticides, including linuron, with accumulation of the corresponding aniline metabolites in the culture media was induced by the herbicide linuron (Engelhardt et al., 1971). The enzyme was not produced in cultures without linuron or in cultures with linuron and a protein inhibitor. Other phenylamides, propanil, propham, 2-chlorobenzanilide, and 2,5-dimethylfuran-3-carboxanilide, induced the enzyme, but with only 6 to 40% of the activity obtained with linuron as the inducer (Engelhardt et al., 1973). Similar amidase activity in whole cells and extracts of *Fusarium oxysporum* was described by Reichel et al. (1991). Linuron and other phenylurea and carbamate herbicides induced cells to degrade propanil, but also inhibited propanil hydrolysis when added to the assay mixture. The phenylurea herbicides were not substrates of the propanil-degrading enzyme. Linuron-induced cells had more activity than when propanil was the inducer. These two systems suggest that enzymes degrading phenylureas, carbamates, and acetanilides are likely to have a range of both inducers and substrates. This is likely to be of practical importance in situations where related compounds are applied together or sequentially, thus creating a preinduced microbial population.

The effects of induction by secondary compounds can be manifested in soils. A *Pseudomonas* which was induced by 4-chloroaniline and 3,4-dichloroaniline degraded both of these compounds as a source of C and N (Zeyer and Kearney, 1982b). If the herbicide propanil (^{14}C-ring-labeled) was added with the *Pseudomonas*, approximately 25% of the propanil ring was mineralized in 4 weeks. If the bacterium was added with unlabeled 4-chloroaniline, ^{14}C-propanil mineralization increased to over 50%, suggesting that there were greater numbers of induced cells present to mineralize the 3,4-dichloroaniline normally produced in the degradation of propanil. Propanil mineralization from non-sterile soil was minimal without the bacterium.

Limited evidence also suggests that the pesticide-degrading enzymes are regulated under a mechanism that is responsive to the overall metabolic status of the cell. The specific activity of the amidase enzyme from *Bacilus sphaericus* grown in media amended with succinate, citrate and glutamate was much lower than activity in minimal media, suggesting regulation analogous to catabolite repression (Engelhardt et al., 1973). *Pseudomonas cepacia* (strain AC1100), which degrades 2,4,5-T and 2,4,5-trichlorophenol (2,4,5-TCP), grown on organic acids prior to the assay failed to dechlorinate 2,4,5-T or 2,4,5-TCP (Karns et al., 1983). Addition of 2,4,5-TCP to succinate media restored activity against 2,4,5-TCP after a lag period showing that chlorinated phenols induced this enzyme. However, the conversion of 2,4,5-T to 2,4,5-TCP was expressed in succinate-grown cells, indicating that this enzyme was constitutively expressed.

C. Multispecies Interactions

Biotransformation of a pesticide by one species can be linked with further reactions by other organisms eventually resulting in the complete degradation of the pesticide. The chloro-*s*-triazine herbicides, such as atrazine, appear to be degraded in this complex manner (Figure 3). The initial steps in the degradation of atrazine results in *N*-dealkylated metabolites, which have been found in soils, subsoils and groundwater (Khan and Saidak, 1981; Schiavon, 1988; Adams and Thurman, 1991; Winkelmann and Klaine, 1991). Bacteria have been isolated which perfom these dealkylations (Behki and Khan, 1986). The propyl and ethyl groups resulting from atrazine dealkylation supported the growth of *Pseudomonas* spp. in carbon-free media, but the triazine ring was not metabolized. Two of the bacteria described by Behki and Khan (1986) could also dechlorinate deethylatrazine (DEA) and deisopropylatrazine (DIA). The inability of soil populations to degrade the triazine ring is further evidenced by the small quantities of $^{14}CO_2$ produced after addition of [^{14}Cring] atrazine (Wolf and Martin, 1975) although other studies have reported as much as 25% ring mineralization in 180 days (Winkelmann and Klaine, 1991). The metabolism of the *s*-triazine ring by bacteria has been observed under defined N-limited conditions (Cook and Hütter, 1981), although no chloro-*s*-triazine compounds were degraded. These organisms were isolated by enrichment from soils that had previous exposure to triazine herbicides. Thus if *s*-triazine herbicides are dealkylated and dechlorinated by one group of soil microorganisms, the resulting ring (e.g. cyanuric acid) could be cleaved, resulting in liberation of CO_2 and NH_4^+, as was found with melamine (Jutzi et al., 1982). Later, Cook and Hütter (1984) described the degradation of deethylsimazine by two bacterial species in defined media. The initial steps resulted in dechlorination and deamination of deethylsimazine, while the second species degraded the ring structure. Although

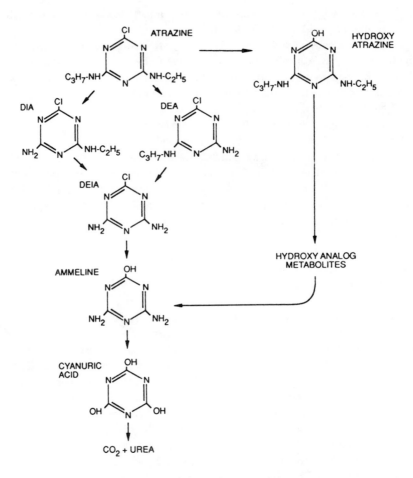

Figure 3. Generalized pathway of atrazine degradation in soil. Atrazine is dealkylated to form DEA and DIA by bacteria and fungi. Conversion of atrazine to hydroxyatrazine by abiotic reactions occurs at increased rates in acidic soils. Some intermediate steps in the conversion of DEDIA to cyanuric acid have been omitted.

these studies demonstrate that microorganisms capable of degrading the s-triazine exist in soil, the relatively low mineralization rate of the triazine ring remains unexplained.

A number of multispecies communities and consortia that degrade organic compounds have been isolated from soils (Slater and Lovatt, 1984). Slater and Lovatt (1984) described communities that are based on the supply of specific growth factors from one member to another, or removal of inhibitory products. Consortia display concerted metabolic action against a complex or recalcitrant

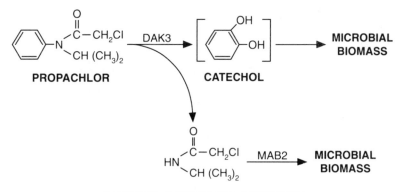

Figure 4. Pathway of propachlor degradation by two soil bacteria (DAK3 and MAB2) isolated from a pesticide disposal site. The metabolite 2-chloror-*N*-isopropylacetamide was produced by DAK3 independently of MAB2 and was identified by GC-MS. See Villarreal et al. (1991) for details.

substrate. Consortia are distinquished from communities by the direct interaction of the members, vs the sequential but independent action of species in communities. This last type of interaction is probably the most common. Propachlor metabolism by a simple two-organism community (Figure 4) was isolated from a pesticide disposal site soil (Villarreal et al., 1991). Strain DAK-3 produced 2-chloro-*N*-isopropylacetamide independently of MAB2. Strain MAB2 did not grow on propachlor, but did metabolize and grow on 2-chloro-*N*-isopropylacetamide. In another study, the herbicide mecoprop was degraded by a stable five-membered community of bacteria isolated from wheat roots (Lappin et al., 1985). The herbicide was utilized as a C source by 5 bacteria cultured together, but individual species were unable to grow on mecoprop and each member was required for complete mecoprop degradation. The metabolites and pathway of degradation was not determined in this study, but chlorine was released into the culture media. The community was isolated using relatively low concentrations of mecoprop (<2.1 mM) after a long period of enrichment.

An example of a more interactive consortium is the co-metabolic degradation of parathion to diethylthiophosphate and *p*-nitrophenol by *Pseudomonas stutzeri*, which was coupled to the degradation of *p*-nitrophenol by *Pseudomonas aeruginosa* (Daughton and Hsieh, 1977). The *P. stutzeri* did not grow on either of the parathion degradation products, and appeared to utilize cellular lysates from *P. aeruginosa*. The original source of the consortium was sewage sludge.

D. Bound Residues

1. Formation

The formation of bound residues is often only considered in the assessment of pesticide fate in soil. In this section an argument will be made for considering this bound residue formation as part of the degradation process. In many cases, bound residue formation is a biological reaction since enzymatic activity is involved. Bound residue formation and transformations in soils are nearly always quantified through the use of ^{14}C-labeled pesticides. Bound residues are experimentally defined as the fraction of pesticide residue (^{14}C) that remains after exhaustive solvent extraction. It is necessary to distinguish between bound residues and strongly sorbed residues (sorption implies that desorption is possible), although this may be difficult experimentally.

Experimental evidence from model systems suggests a generalized mechanism accounting for the formation of non-extractable (bound) residues. Microbial phenoloxidases, laccases and peroxidases react with various pesticides or their metabolites with model organic matter components to form complex polymers with the pesticide residue incorporated by covalent bonding. Bollag et al. (1980) presented evidence indicating that an extracellular laccase enzyme from *Rhizoctonia praticola* catalyzed reactions between 2,4-dichlorophenol and orcinol, syringic and vanillic acids to produce various polymeric products. Similarly, soil or laccase enzyme catalyses the reaction of syringic acid with 2,6-diethylaniline, a potential metabolite of alachlor as shown in Figure 5 (Liu et al., 1987). Autoclaved soil formed no product from these reactants, but gamma irradiated soils partially retained the capability for reaction. This illustrates that both biological and chemical reactions may be involved. The pesticide residue-organic polymer presumably undergoes further reaction to become incorporated within the soil organic matter. Later studies showed that laccase immobilized on clay particles had activity and the clays protected the laccase from proteases (Ruggiero et al., 1989). These findings demonstrate a microbial role, either directly or as a source of extracellular laccase, in bound residue formation and provide several model mechanisms for bound residue formation. Abiotic mechanisms include chemical bonding, hydrophobic partitioning or entrapment within the organic matter polymers. Entrapped residues may undergo nonenzymatic chemical reactions to produce organic matter bonded with pesticide residues.

The evidence obtained from reacting pesticides with model humic acid constituents is complemented by many studies quantifying the incorporation of ^{14}C, originally applied as herbicide, into various organic matter fractions. Experimentally, the bound ^{14}C is considered that which remains in the soil after exhaustive aqueous or solvent extractions. Typically the formation of bound ^{14}C-residues increase steadily during the first weeks after pesticide application, then more slowly until reaching a maximum. Sterilization of soil prior to pesticide addition reduces or eliminates bound residue formation for many compounds.

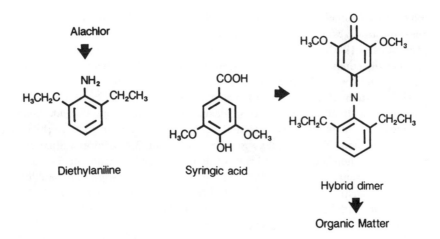

Figure 5. Proposed reaction pathway of diethylaniline, a microbial metabolite of alachlor, with syringic acid in the presence of microbial laccase enzyme. Syringic acid is a naturally-occurring compound which potentially is incorporated into soil organic matter polymers. The diethylaniline-syringic acid dimer could be further reacted with soil organics to form bound pesticide residue. (Based on Liu et al., 1987.)

Total ^{14}C-bound residues typically ranges between 15 and 60% of applied ^{14}C for heterocyclic-ring herbicides such as triazines and dinitroanilines, with lesser amounts from phenoxy- or benzoic-acid herbicides (Khan, 1982; Nelson et al., 1983; Smith and Muir, 1984). These studies show nearly equal distribution of ^{14}C-residues in the humic and fulvic acids, with slightly more in the humin fraction. In other studies, more ^{14}C was found to be associated with the fulvic acid than the other fractions (Capriel et al., 1985; Lee et al., 1988). Certainly, in quantitative terms the formation of bound residues is an important degradative process that needs to be considered in any assessment of pesticide fate.

2. Bioavailability and Fate

The fate of these bound residues is an important question relative to soil quality. Several studies have shown that small amounts of parent pesticide and metabolite can be extracted from bound residues through the use of nonconventional methods. For instance, using a high temperature distillation, Capriel et al. (1985) extracted atrazine and several metabolites from the bound residue fraction 9 years after outdoor application of ^{14}C-atrazine. Of the initial applied radioactivity 46% was still present in the bound fraction. Concentrations of atrazine or metabolites did not exceed 0.2 mg kg^{-1} soil. The recovery of these compounds through various chemical extraction procedures does not necessarily indicate bioavailability. Maize plants took up only 0.7% of bound ^{14}C-metha-benzthiazuron residues, which accounted for 41% of applied ^{14}C (Führ and Mittelstaedt, 1980). Other studies also show very low plant uptake of bound residues in comparison to either aged or freshly added herbicide residues (Helling and Krivonak, 1978; Lee et al., 1988).

The question of availability of bound residues to soil microorganisms has received less attention. Soil microorganisms can transform and degrade bound residues. Products of these reactions can include the original herbicide, metabolites, and CO_2. Complexes of humic acid with 3,4-dichloroaniline were mineralized when added back to soil, while in sterilized soils no mineralization occurred (Saxena and Bartha, 1983). Mineralization rates were comparable to that of the native soil organic matter. Further evidence was supplied in experiments with bound parathion residues, in which 17% of the bound residue was mineralized after mixing with fresh soil (Racke and Lichtenstein, 1985). Mixing the bound residues with cow manure and soil increased the amount mineralized to 27% and stimulated microbial populations. Bound residue uptake by oat plants in these treatments was not affected by the cow manure. Studies with bound residues of prometryne showed that different physiological groups of microorganisms (cellulolytic, proteolytic, lipolytic and lignolytic groups) did not differ in their abliity to release and mineralize bound ^{14}C-residues (Khan and Ivarson, 1982). Later studies showed that release of atrazine and metabolites increased after inoculation of atrazine-degrading bacteria compared to the indigenous microorganisms (Khan and Behki, 1990). Greater amounts of DIA and hydroxydeisopropylatrazine were recovered as incubation time increased. It was not determined if these metabolites were released directly from the bound residue or resulted from metabolism of atrazine after release from the organic matter.

A two-step process is suggested, first the enzymatic cleavage of the organic matter-pesticide residue bond, then the biodegradation of the pesticide residue. The rates of this process are slow, since bound-residues behave more like soil organic matter than freshly added pesticides. Bound residues appear to be largely unavailable for leaching or plant uptake. However, repeated application of pesticides in the field would lead to a potentially large pool of bound materials that would represent a mixture of parent pesticide and metabolites. If only a small fraction of this bound residue were available it could be a potential contributing factor to leaching of pesticides.

III. Rates of Biodegradation

A. Kinetics of Biodegradation

Estimates of biodegradation rates are necessary for predicting pesticide behavior in soil. The development of increasingly sophisticated models reflects the need for predicting herbicide performance and environmental impact. Accurate descriptions of biodegradation kinetics may improve these simulation models. Kinetic models also serve the useful purpose of integrating the physiological activity and population dynamics of microorganisms degrading a particular compound. In many practical situations, such as the bioremediation of pesticide-contaminated soil, kinetic models may have some diagnostic value in determining if particular biological degradation processes are rate-limited. Considerable development may be required to use these kinetic models in practical applications. Pesticides which are mineralized slowly or degraded by a consortium of microorganisms may require simplifying assumptions to be made or the use of alternative kinetic models. The development of accurate and rapid methods for quantifying pesticide-degrading microorganisms would allow verification of the Monod models which require an explicit estimate of degrader biomass.

1. Monod Kinetics

The degradation of substrates by microorganisms follow saturation kinetics of the Monod type (Table 1). The Monod model couples bacterial growth and substrate utilization over a broad range of substrate concentrations. Theoretically, Monod kinetics apply to the utilization of single substrates by homogeneous populations of microorganisms. Variants of this model, based on the relative size of the biomass and substrate concentration, include logistic, logarithmic, zero-order and first-order kinetics (Alexander and Scow, 1989). For instance, the Monod model reduces to the first-order model when the substrate affinity parameter, K_s, is much larger than the substrate concentration. Measurements of mineralization and biomass of specific degraders, or incorporation of substrate into biomass are required for model fitting.

The Monod model has been applied to the degradation of a broad range of simple sugars and unsubstituted aromatic substrates in cultural media or in freshwater samples (Simkins and Alexander, 1984; Simkins and Alexander, 1985; Jones and Alexander, 1986). Greer et al. (1992) examined the differences in degradation kinetics of seven 2,4-D degrading bacteria in culture, including species of *Pseudomonas*, *Alcaligenes*, and *Bordetella*. Estimates of K_s ranged from 2.2 to 33.8 mg L^{-1}. This suggests substantial diversity. Soil solution concentrations of 2,4-D would be expected to be below these K_s values. Doubling times for these strains were less variable, ranging from 3.5 to 2.2 hr. Larson (1980) reported that Monod kinetics described the mineralization of two quartenary surfactants in freshwater samples containing natural heterogeneous bacterial populations. Jones and Alexander (1986) attempted to describe mineralization of low concentrations of phenol in natural lake-water. Model descriptions of the kinetics were not completely consistent over time; it was suggested that temporal dynamics of the indigenous populations and the lake water chemistry were important.

Table 1. Equations governing the kinetics of biodegradation of organic compounds by microorganisms

Model	Equation Form	Restrictions
Monod (with growth)	$-dS/dt = \mu_m SB/Y(k_s + S)$	
Monrod (no growth)	$-dS/dt = v_m S/(k_s + S)$	B is constant
Zero-order	$-dS/dt = k_0$	$S >> K_s,\ B >> S$
First-order	$-dS/dt = k_1 S$	$k_1 \cong \mu_m B/Yk_s$ with $k_s >> S$ B is constant
Second-order	$-dS/dt = k_2 SB$	$k_2 \cong \mu m/YK_s$ with $k_s >> S$

Variables: S: substrate (pesticide) concentration, B: biomass of degrading population, μ_m: maximum microbial growth rate on substrate (S), v_m: maximum degradation rate per unit of biomass or enzyme, Y: yield factor (conversion of S into B), k_s: substrate affinity parameter (S at $\frac{1}{2}\mu_m$). Parameters k_0, k_1, and k_2 are rate constants as defined above. See Paris et al., 1981; Simkins and Alexander, 1984; Alexander and Scow, 1989 for further details.

Modified forms of Monod kinetics have been applied to the degradation of a few aromatic organic compounds in soil. In a forest soil treated with 0.1 mg kg^{-1} soil of dinitrophenol, degradation followed first-order kinetics, while at higher concentrations Monod or dual-substrate (a modification of Monod kinetics allowing for secondary substrates) models provided better descriptions of the data (Schmidt and Gier, 1989). Mineralization curves at these higher concentrations were generally sigmoid in appearance. Based on model results and most-probable-number (MPN) estimates of populations of dinitrophenol-degraders, the threshold concentration supporting population growth above the indigenous population of 3×10^4 cells g^{-1} soil was greater than 10 mg kg^{-1} soil of dinitrophenol. Later studies (Schmidt and Gier, 1990) isolated dinitrophenol-degrading species of *Janthinobacterium* and *Rhodoccocus* from the same soil. The more numerous *Janthinobacterium* had lower k_s and μ_m than *Rhodococcus*, suggesting that *Janthinobacterium* was responsible for mineralization at lower concentrations of dinitrophenol. At higher concentrations of dinitrophenol the *Rhodococcus* would probably become increasingly more important.

In other studies with soil Scow et al. (1986) reported that mineralization of several phenolic compounds and other simple aromatic organics were not described by Monod-type models, but that a version of two-compartment kinetics did describe the mineralization process.

2. First-order Kinetics

First-order kinetics have often been used to describe pesticide degradation or dissipation in soil. First-order kinetics assume that the rate of degradation is proportional to the concentration of pesticide, within certain limits. Implicit in the acceptance of first-order kinetics are the additional assumptions that the biomass degrading the pesticide does not change (at least within the time-course of the degradation process) and that the substrate concentration (pesticide) is the limiting factor. Within certain concentration ranges these assumptions appear to hold true for many chemicals in many types of soils, as determined by least-squares regression analyses of pesticide residue data. Nash (1988) has compiled an extensive review of degradation rate constants estimated using first-order kinetics for many pesticides under field and laboratory conditions.

While the first-order model appears to describe degradation reasonably well in individual soils, certain problems remain unresolved. The range of concentrations that are described by a rate constant appear to be narrow. For instance, rate constants describing fluometuron degradation at 1500 ng g^{-1} soil did not describe degradation at 85 ng g^{-1} (Mueller et al., 1992), showing that the process was not true first-order. In some studies first-order kinetics do not adequately describe the shape of the degradation curve. The reaction order for metribuzin degradation was greater than 1.5 in 14 of 18 soils in laboratory incubations (Allen and Walker, 1987). Alternative approaches have been proposed including two-compartment models (eg. Hill and Schaalje, 1985) and multi-compartment approaches (Gustafson and Holden, 1990). Hill and Schaalje (1985) proposed a two-compartment model to describe the rapid initial dissipation and subsequent slow phase observed in field dissipation studies. Locke and Harper (1991a) used a similar two-compartment model to account for metribuzin degradation in laboratory systems. They interpreted the slow phase to be degradation of strongly sorbed herbicide that was only slowly desorbing back into the soil solution.

The first-order rate constant describing degradation in one soil rarely describes degradation in another soil, even in laboratory incubations where environmental effects and competing processes are minimized. Laboratory incubation experiments (uniform temperature and soil water potential) with nine agricultural soils resulted in half-lives for napropamide ranging from 72-150 days (Walker et al., 1985). Adsorption accounted for some variability ($r^2 = 0.45$), but the unexplained variation in degradation rates was attributed to differences in microbiological activity or biomass. Fenamiphos half-lives ranged from 12 to 87 days (coefficient of variation of 49%) in a similar study with 16 diverse soils (Simon et al., 1992). In later studies comparing degradation in low organic matter soils of neutral pH (6.3 to 7.4) under uniform temperature and moisture conditions, metamitron and metazachlor degradation rates had coefficients of variability of 17% and 10%, respectively (Allen and Walker, 1987). Correlations obtained with soil properties were generally low ($r < 0.62$), although metamitron degradation appeared to be influenced by soil factors controlling herbicide availability. The variability in degradation rate constants of pesticides in different soils that cannot be attributed to soil properties suggests variability in the microbiological capacity to degrade these compounds.

Apparent first-order kinetics have often been used to describe pesticide dissipation under field conditions. In the field, the processes of leaching and volatilization contribute to the dissipation of applied pesticide, as determined from soil residue measurements. In laboratory systems where these other processes can be minimized or eliminated, degradation rates can be measured more accurately. Frehse and Anderson (1983) correlated field and laboratory half-lives for 27 pesticides; for most compounds degradation was faster under field conditions, probably reflecting the losses due to volatilization and leaching.

3. Second-order Kinetics

Second-order kinetics can be derived from Monod kinetics, but require no knowledge of yield factors or other parameters. The most straightforward approach to the incorporation of microbial biomass into degradation kinetics involves normalizing the degradation rate on a biomass basis, or second-order kinetics, as shown in Table 2 (Baughman et al., 1980; Larson, 1984). Second-order kinetics have been used with some success in aquatic systems. Paris et al. (1981) examined the hydrolysis rates of esters of 2,4-D and malathion. Populations of heterotrophic bacteria were used to obtain fairly consistent second-order rate constants. In experiments where the size of the microbial biomass was manipulated in a particular parabrown soil, the amount of diallate and triallate degraded (the sum of that mineralized and converted to bound residue) was proportional to the microbial biomass present in the soil (Anderson, 1984), which appears to support the concept of second-order kinetics. Mueller et al. (1992) found that specific rate constants (first-order rate constant/unit of microbial biomass) ranged from 1.37×10^{-4} to 1.59×10^{-4} in surface soils (0-30 cm) and from 0.68×10^{-4} to 0.34×10^{-4} in subsurface soils (30-120 cm depth). The range of these specific rate constants indicates some deviation from strict second-order kinetics. In both of these examples total microbial biomass was used as a surrogate for populations of specific degraders. For some pesticides, where the specific degrader population is a large or constant fraction of the total population this may be a useful approach. If second-order kinetics were shown to be generally applicable to pesticide degradation in soils, more accurate estimates of pesticide degradation could be made over different soils from a generalized second-order rate constant and measurements of the microbial biomass. More complex interactions are also possible, such as the correlation between fenamiphos mineralization and the ratio of microbial biomass C to total soil organic C (Simon et al., 1992). These latter observations require further study to determine their significance.

Table 2. Populations of pesticide-degraders in soils in comparison to total heterotrophic bacterial population. Populations are restricted to those in soils without enhanced rates of biodegradation and are based on the ^{14}C-MPN technique as an enumeration method. Letters in the parentheses indicate source of data

| Soil | Pesticide | Population (cells g^{-1} soil) | |
		Total	Degraders
Cecil	2,4-D	1.7×10^7	2.4×10^2 (a)
Webster	2,4-D	3.9×10^7	9.3×10^2 (a)
Indian HeadTama	2,4-D	1.7×10^7	25×10^2 (b)
Drummer	isofenphos	1.2×10^8	ND (c)
Norfolk	carbofuran	8×10^6	2.1×10^5 (d)
Granada	carbofuran	1.35×10^7	1.6×10^3 (e)
	EPTC	1×10^7	1.8×10^6 (f)

ND: below detectable levels.
[a]Ou, 1984; [b]Cullimore, 1981; [c]Racke and Coats, 1987; [d]Dzantor and Felsot, 1989; [e]Hendry and Richardson, 1988; [f]Moorman, 1988.

B. Populations of Biodegrading Microorganisms

Little information is available concerning the natural populations or biomass of microorganisms responsible for degrading pesticides. Knowledge in this area is limited by the lack of reliable methods for enumerating degraders of many pesticides. Pesticide-degrading populations appear to be a small fraction of the total population for many pesticides as shown in Table 2, although EPTC-degrader populations approach 20% of the total population. The populations shown in Table 2 represent those organisms capable of ring mineralization (except for EPTC), and do not necessarily reflect the contribution of other populations which biotransform without mineralization.

Several studies have shown that microbial growth from pesticide degradation is most evident as the pesticide application rate increases. For instance, Nelson et al. (1982) showed that increases in populations of parathion-degrading bacteria were proportional to the rates of application. The greatest increase was from 25×10^5 to 300×10^5 cells g^{-1} soil during the 6-day period after application of 160 mg kg^{-1} soil of parathion, but the population declined afterwards. It was suggested that the decline was due to toxic accumulations of the metabolite p-nitrophenol. Little is known concerning the longer-term temporal or spatial variability of populations of pesticide-degrading microorganisms.

C. Adaptation and Enhanced Biodegradation

1. Adaptation

The adaptation phenomenon refers to the increase in microbial degradation of a compound resulting from exposure to the compound. Adaptive responses have been observed in pure cultures, aquatic environments, soils, and subsurface aquifers (Moorman, 1990). Adaptation has been used to describe shortened lag-periods preceding degradation or increases in degradation rates in adapted systems compared to non-adapted systems, developing in response to single or repeated exposures. Both short-term and long-term adaptive responses have been observed and the diversity of the responses suggests that different mechanisms are involved. The processes that contribute to adaptation include enzyme synthesis, population growth, selection and/or genetic transfer or some combination of these responses.

2. Enhanced Biodegradation

Enhanced or accelerated biodegradation has been reported to develop in response to agricultural applications of pesticides at typical frequencies and rates. The rapid degradation of pesticides by adapted soil microbial populations has resulted in reduced pest control and yield (Roeth, 1986; Felsot, 1989). With respect to herbicides, this phenomenon was first reported by Audus (1949) for the biodegradation of 2,4-D. Carbamothioate and carbamate herbicides appear to be most susceptible to this phenomenon, but many classes of herbicides and insecticides are affected. In addition to carbamothioates, enhanced biodegradation has been reported for the carbamate insecticide carbofuran (Felsot et al., 1981; Read, 1983; Turco and Konopka, 1990), the dicarboxamide fungicides iprodione and vinclozolin (Walker, 1987), and organophosphorus insecticides diazinon (Forrest et al., 1981), fensulfothion (Read, 1983) and isfoenphos (Racke and Coats, 1987). This represents only a partial list featuring compounds that have demonstrated problems related to efficacy.

Carbamothioates, such as EPTC and butylate, are more rapidly mineralized in soils that have been repeatedly treated with these herbicides than in soils without previous exposure to carbamothioates (eg. Obrigawitch et al., 1982; Skipper et al., 1986). At least two previous annual applications can cause adaptation. Studies with ^{14}C-labeled herbicides show that significant amounts of herbicide are rapidly mineralized, which also indicates a high degree of microbial involvement. Commonly, mineralization curves such as those in Figure 6 are observed. The lower initial rates are followed by rapid rates of mineralization. The duration of this initial lag phase is less in the soils with adapted populations than in soils with nonadapted populations (Obrigawitch et al., 1982; Skipper et al., 1986; Moorman et al., 1992). The initial lag phase can be interpreted as time required for enzyme induction and synthesis. The

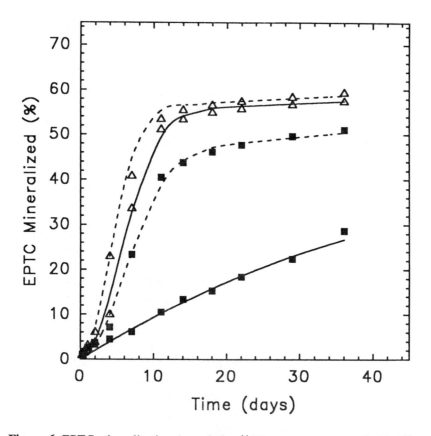

Figure 6. EPTC mineralization (cumulative $^{14}CO_2$ as a percentage of added ^{14}C) in Grenada silt loam soil with 6-year previous EPTC use (dotted lines) in comparison to unexposed soil (solid lines). EPTC was added as [^{14}C-*carbonyl*] EPTC (open triangles) or [^{14}C-*N-propyl*] EPTC (closed squares). Soils were incubated under uniform water potential and temperature conditions. Symbols indicate means and lines indicate a fitted regression line. (See Moorman et al., 1992 for details.)

involvement of microorganisms is further indicated by the effectiveness of various sterilization and broad spectrum inhibitory agents in eliminating or decreasing the degradation of the herbicides. Antifungal agents were less effective inhibitors of EPTC mineralization than inhibitors of bacteria, suggesting that bacteria were the adapted microorganisms (Tal et al., 1989). Both adapted and nonadapted soils are inhibited to the same extent.

Microorganisms that degrade carbamothioates have also been isolated. Lee (1984) reported that a several genera of soil fungi and bacteria isolated from enrichments degraded EPTC, but noted that the ability to degrade EPTC was lost after 15 months of storage. Species of *Rhodococcus* (Dick et al., 1990), *Arthrobacter* (Tam et al. 1987), and *Flavobacterium* (Mueller et al.,1988) isolated from soil utilized carbamothioates as sole C sources. Tal et al. (1990) also isolated a bacterium which grew on EPTC, and two other carbamothioate herbicides, butylate and vernolate, as C sources. This isolate inoculated at high levels (10^7 cells g^{-1} soil) increased the rate of EPTC degradation in soil. Loss of a plasmid from the *Arthrobacter* coincided with loss of degradative ability.

The ability of microbial populations with enhanced degradation rates relative to one pesticide to degrade a structurally similar molecule has been termed cross-adaptation. For example, vernolate and butylate were degraded more quickly in a soil with a microbial population adapted to EPTC than in a soil unexposed to EPTC (Obrigawitch et al., 1983). In some instances, the cross-adaptation has been reported with molecules that are not structurally similar. Soils adapted to the carbamothioate herbicide triallate degraded EPTC at enhanced rates, but also degraded the unrelated insecticide, carbofuran at greater rates (Cotterill and Owen, 1989). Aldicarb degradation was unaffected by previous triallate applications.

The isolation of microorganisms capable of degrading and utilizing carbamo-thioate herbicides establishes population growth as one possible mechanism accounting for enhanced biodegradation rates in soils. Mueller et al. (1989), using a TTC indicator medium, found that populations of actinomycetes and other bacteria degrading carbamothioate herbicides in soils with histories of EPTC use were increased 6 weeks after herbicide application. Some populations, such as EPTC-degrading actinomycetes did not increase in the soil following EPTC use. It was not determined if these enlarged populations of degraders persisted throughout a full cropping cycle. Elsewhere, Moorman (1988) showed that enhanced and nonenhanced soils did not differ in their spring-time, preapplication populations of EPTC-degraders. Less than 3% of applied ^{14}C-EPTC was incorporated into the microbial biomass in these soils, which supports the lack of population response to prior EPTC applications (Moorman et al., 1992). As the EPTC application rate increased, degrader populations increased in the adapted soil during laboratory incubations (Figure 7).

Different results have been obtained with other pesticides with respect to population responses and adaptation. Carbofuran-degraders were increased after two 10 mg kg^{-1} soil applications of carbofuran spaced 10 months apart, compared to untreated soils (Dzantor and Felsot, 1989). These population increases corresponded to increases in carbofuran degradation. Increases in isofenphos-degraders also corresponded to the development of enhanced degradation (Racke and Coats, 1987), but comparisons between adapted and nonadapted soils of the same soil series were not made. Populations in non-adapted soils were essentially nondetectable, but in isofenphos-adapted soils, populations ranged from 6×10^3 to 12×10^3 degraders g^{-1} soil. Mineralization of [^{14}C-ring] isofenphos was also increased in isofenphos-adapted soils and isopropyl salicilate was extracted as a metabolite in soils and microbial cultures. Population growth appears to play some role in the adaptation to carbofuran and

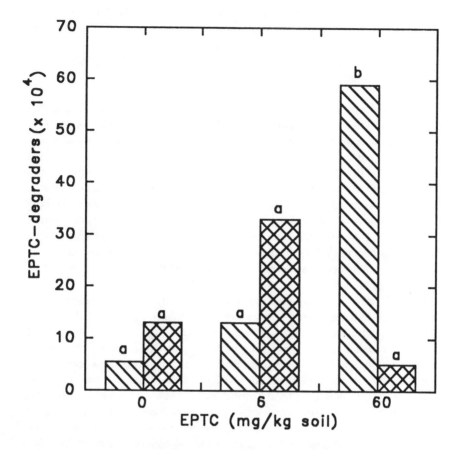

Figure 7. Populations of EPTC-degrading bacteria in Grenada silt loam soil with previous history of EPTC use (diagonal shading) and without prior EPTC use (crosshatch shading). Populations were determined by a ^{14}C-MPN procedure on soils after incubation for 25 days after amendment with different amounts of EPTC. Letters indicate significant differences. (From Moorman, 1988.)

isofenphos, although the potential growth on C released in pesticide metabolism appears to be limited due to the low amount of C in pesticide applications and the maintenance requirements of actively growing populations (Moorman, 1990).

An alternative or additional hypothesis to population growth as an explanation for microbial adaptation to pesticides is the possibility that microorganisms have developed more efficient pathways of degradation. Experiments with the same soils used for studies of EPTC-degrader populations (Moorman et al., 1992) compared mineralization of [^{14}C-carbonyl] and [[14]C-*N*-propyl] EPTC (Figure 6). The initial rates of mineralization were low in both adapted and nonadapted soils, but the lag phase was of shorter duration on the adapted soil (Moorman et al., 1992). Adaptation was greatest in the metabolism of the *N*-propyl portion of the molecule, although mineralization of [^{14}C-carbonyl] EPTC was also enhanced in the adapted soil (Figure 6). The evidence suggests that microorganisms capable of degrading the *N*-propyl portion of the molecule, such as the *Rhodococcus* isolated by Dick et al. (1990), have become enriched in the adapted soil.

Other evidence suggests that pesticide metabolites enhance the degradative capability of the soil community. Pretreatment with a metabolite before [^{14}C-ring]2,4-D addition increased $^{14}CO_2$ evolution from 70% in untreated soil to 90% in soils treated 4 times with 500 mg kg^{-1} of 2,4-dichlorophenol (Somasundaram et al., 1989). In similar experiments, pretreatment with *p*-nitrophenol enhanced parathion degradation and salicilic acid enhanced degradation of isofenphos. Although the metabolites may induce microbial enzymes, the large concentrations of metabolites used in the pretreatments could support considerable microbial growth. In fields receiving recommended pesticide applications, metabolites may not persist from one year to the next or may be present only at trace levels. Whether trace levels of these compounds can induce degradative systems is unknown.

Plasmids have been suggested as a potential factor in the development of adapted soils. Carbamothioate-degrading bacteria have plasmids that appear to carry degradative genes; plasmid-less strains do not degrade herbicides (Tam et al, 1987; Mueller et al., 1988). The herbicide 2,4-D, which is also prone to enhanced biodegradation, is degraded by microorganisms with plasmid-encoded degradative genes. Holben et al. (1992) examined populations of 2,4-D degraders in soils with long-term histories of 2,4-D use and found no difference in the populations of exposed and non-exposed soils. Laboratory amendments of 250 mg kg^{-1} soil of 2,4-D caused dramatic population increases, as determined by MPN. Gene probes to 2,4-D oxygenase and 2,4-dichlorophenol hydroxylase from the pJP4 plasmid generally showed the same trends as the MPN populations. Although plasmid transfer in soils has been demonstrated with a variety of species, there is no strong evidence suggesting or discounting plasmid transfer as a factor in the short (2 to 3 growing seasons) times required for development of enhanced biodegradation. Evidence for plasmid transfer as a factor in the development of enhanced biodegradation would be demonstration of an increase in plasmid frequency in response to pesticide use history and concomitant enhancement of degradation. Plasmid transfer of degradative abilities in response to some selective mechanism other than pesticide applications may be possible.

IV. Environmental Factors Affecting Biodegradation in Soil

A. Physical Limitations

Availability of pesticide in the soil environment can affect the magnitude and rate of biodegradation. Sorption processes control the distribution of pesticides between the soil solution and soil particles. Sorption partition coefficients (K_d, K_f, K_{oc}) describe the distribution of herbicide between soil and water at equilibrium or near-equilibrium conditions. The nature of the sorption process is only partly understood; electrostatic interactions, hydrogen bonding, and hydrophobic partitioning are some of the mechanisms known to participate in the sorption process (Hassett and Banwart, 1989). Theoretically, partition coefficients represent the distribution of the chemical at equilibrium, which implies that the sorbed fraction can be desorbed. Pesticides bound by chemisorptive mechanisms (covalently bonded compounds, which are functionally equivalent to bound residues) are excluded from the equilibrium process and are considered unavailable for degradation, except as part of the soil organic matter.

1. Availability of Adsorbed Pesticides

Adsorbed pesticides are often considered to be less available for biodegradation than chemicals in solution. If sorbed pesticides occupy sites that do not allow direct contact with microorganisms, then molecules must desorb from soil surfaces and move by diffusion to cell surfaces for uptake to occur. When the rate of biodegradation exceeds the rate of desorption, then desorption becomes the rate-limiting step. Diffusional processes are implicitly included in both the desorption and degradation processes, when these processes are measured independently using conventional batch-reaction techniques. The exact nature of the sites protecting pesticides from degradation are not known, but they could include spaces between clay platelets, micropores within mineral particles or regions within soil organic material.

For purposes of initial assessment, it appears that pesticide degradation rates in soil are far slower than rates of adsorption and desorption. For instance, sorption equilibria of moderately soluble pesticides such as atrazine, alachlor and fluometuron are such that a significant fraction of the herbicide remains in solution at equilibrium, and a near-equilibrium is established within 24 hr (Locke, 1992; Mueller et al., 1992). The half-lives of these compounds are far greater than the time necessary for equilibrium to establish. If desorption rates are roughly equivalent to degradation rates, the solution phase concentration can be reestablished as the concentration of pesticide is decreased. Nevertheless, the rate of biodegradation of some compounds is decreased in the presence of a sorbent (Gordon and Millero, 1985; Ogram et al., 1985; Wszolek and Alexander, 1979). Degradation of a series of n-alkylamines decreased as their sorption to bentonite increased, relative to degradation in bentonite-free medium (Wszolek and Alexander, 1979). However, as the size of the population of bacteria degrading n-decylamine was increased to very high levels, the rate of biodegradation appeared to exceed the maximum rates of desorption, which was attributed to bacterial modification of the sorption equilibrium through pH changes or biosurfactant production. Simultaneous measurements of sorbed and solution concentrations of carbofuran showed that solution phase carbofuran was decreased at a greater rate, suggesting that the rate of desorption was limiting degradation (Shelton and Parkin, 1991). Ogram et al. (1985) found that mineralization of the weakly-sorbed herbicide, 2,4-D was best described by a model that assumed adsorbed herbicide to be unavailable for degradation, but allowed for solution-phase 2,4-D degradation by bacterial cells adsorbed to soil surfaces and in solution.

2. Strongly Adsorbed and Protected Fractions

Some evidence suggests that the assumptions of equilibrium in sorption processes do not always apply to the desorption of aged residues. It has been suggested that a certain fraction of sorbed residues becomes highly resistant to desorption, either by strong chemical binding or by diffusional barriers at microscopic or molecular scales (Hamaker and Goring, 1976; Pignatello, 1989). This fraction is protected from degradation by physical exclusion of the microorganisms. Two-compartment models partition pesticides into sorbed pesticide which is readily desorbed (labile) and pesticide which is only slowly available to the labile or solution pools. Microorganisms remove pesticide from the solution phase only, but desorption replenishes the solution concentration as degradation proceeds. Plots of pesticide remaining vs time for pesticides undergoing degradation under these constraints show an initial rapid phase of degradation followed by a slower phase. It is thought that the second, slower phase is controlled by the movement of herbicide out of the protected compartment after the labile compartment concentration has been depleted.

These types of kinetics have been applied to the degradation of herbicides triclopyr (Hamaker and Goring, 1976) and picloram (McCall and Agin, 1985). In the case of picloram, the partition coefficients which included labile and protected picloram in the adsorbed phase increased substantially during 100 d in soil. Thus, as the herbicides aged, the rapidly desorbable fraction was decreased by degradation and the protected fraction increased. The aged picloram residues desorbed at much lower rates than freshly added picloram. Steinberg et al. (1987) compared the degradation of previously added ethylene dibromide (EDB) to that of freshly added EDB. The aged EDB residues were protected from biodegradation by their close association to the soil. EDB added to the same soils was degraded immediately. Physical disruption caused release of the EDB in the protected fraction, which supports the concept of physical protection from biodegradation.

Short-term evaluations of sorption kinetics also provide evidence for the existence of a protected fraction of sorbed chemical. Locke (1992) described the sorption and desorption of ^{14}C-alachlor over a 96-hr period using a model that partitioned the sorbed fraction into labile and restricted (protected) sites. As equilibration time increased, the model predicted an increase in the nonlabile fraction. Between 9 and 13% of the ^{14}C was methanol-extractable, but not desorbed by 0.01 M $CaCl_2$. Presumably, this corresponds to the herbicide in the restricted sites. In the same studies additional herbicide was not extracted by methanol, and may represent a bound residue that is unavailable for desorption. Short-chain chlorinated aliphatic compounds (trichloroethylene and others) also formed nondesorbable fractions during sorption experiments, which accounted for 10% or less of the applied compound (Pignatello, 1990a, 1990b). The concentrations of these compounds were greatest in particles 250-2000 μm in diameter. The compounds were released by physical disruption.

Soil aggregates may also provide a physical limitation to biodegradation by slowing diffusion of pesticides to cells. Scow and Hutson (1992) developed a model (DSB) that accounts for sorption and diffusion in and out of aggregates. Model simulations indicated that aggregate size (0.1 to 1 cm radius) and sorption interacted. Biodegradation was slowest under the conditions of larger aggregate size and greater sorption. The effects of sorption and diffusion were more pronounced at lower rates of degradation, and degradation was assumed to occur only in the solution external to the aggregates (Scow and Hutson, 1992). Using model systems of clay or polyacrylamide aggregates and a *Pseudomonas* sp., Scow and Alexander (1992) tested the DSB model for a variety of organic substrates. In most of the substrate-aggregate combinations the model provided a reasonable prediction of biodegradation. These models and simple experimental systems illustrate the potential effects of sorption and diffusion on the biodegradation process. The DSB model does not describe the existence of labile and protected fractions of sorbed chemical, although intraaggregate concentrations are protected from degradation. Soil aggregates which have intraaggregate populations of microorganisms may have degradation occurring within the aggregates.

B. Temperature and Moisture Effects

Soil temperature and moisture affect several processes that contribute to dissipation (loss of pesticide from a certain volume of soil) under field conditions. Temperature and moisture effects on microbial activity affect biodegradation, but the volatilization and transport processes are also affected. Within the range of temperature conditions normally encountered in cultivated soils (0 to 35°C) the rate of degradation generally increases with temperature and moisture content. The temperature dependence of degradation rate constants has been described (see Nash, 1988 for details) by a modified Arrhenius equation:

$$k_t = A_0 \, (e^{-\Delta E/RT})$$

where k_t is the degradation rate constant at some temperature (T), A_0 is a constant, and ΔE is the Arrhenius activation energy. The activation energy calculated for a number of herbicides was compiled by Nash (1988). For individual herbicides in different soils, activation energies varied by 20% in most cases, although some estimates differed markedly from others. For instance, estimates of activation energy for amidosulfuron in 3 Canadian soils were 23, 58, and 41 kJ mol[-1] (Smith and Aubin, 1992). Variability in the activation energies may reflect different temperature response profiles of the microorganisms in soils and the effect of temperature on sorption and diffusion.

Low soil moisture contents affect pesticide degradation by reducing microbial biomass and activity and by reducing the total pesticide available in the soil solution (Anderson, 1981). As water potential decreases, soil water films will become more discontinuous and the diffusion of herbicide to microbial cells will become more difficult. Shelton and Parkin (1991) concluded that the principal effects of low soil moisture were on microbial activity, rather than pesticide bioavailability. These conclusions were based on comparative studies of sorbed and solution carbofuran concentrations in soil under several water potentials. Metabolism reduced solution-phase carbofuran rapidly in -0.4 bar water potential soil resulting in an increase in adsorption. This effect was not seen in soil at -7 bars due to the reduced microbial activity. Walker and Barnes (1981) devised an empirical relationship to describe the effect of soil moisture content on degradation rates:

$$H = AM^{-B}$$

where A and B are constants and M is the soil moisture content, and H is the half-life. Like the activation energy, estimates of B generally agree for a particular pesticide, but with some variability (Nash, 1988).

C. Soil Depth

Pesticides generally become more persistent with increasing depth in the soil. This trend has been observed for a wide variety of compounds including the herbicides metribuzin (Kempson-Jones and Hance, 1979; Moorman and Harper, 1989), alachlor (Pothuluri et al., 1990) and fluometuron (Mueller et al., 1992). The rate of degradation changes over relatively short distances. For instance, fluometuron half-lives increased from approximately 20 days in the surface 15 cm of soil to 147 days in soil from the 90 to 120 cm depth (Mueller et al., 1992). Significant positive correlations were obtained with respiration and microbial biomass, which also declined with depth. The low rates of degradation in shallow subsoils would appear to provide increased opportunity for further leaching of compounds escaping from the surface layer.

Aquifer microorganisms often have surprisingly large populations, generally ranging between 10^4 and 10^6 cells g[-1] sediment (Ghiorse and Wilson, 1988). Pesticide degradation in the deeper subsurface may be limited by the low populations of degrading strains, or by environmental conditions such as nutrient limitations. Experiments with alachlor showed that nutrient additions stimulated degradation in subsurface materials, but the effect was not consistent at all depths (Pothuluri et al., 1990). The rates of alachlor degradation in the aquifer samples (12-15 m depth) were very low, with estimated half-lives exceeding 300 days. Other studies with alachlor showed metabolism without significant mineralization by groundwater microorganisms (Novick et al., 1986) Other studies with metolachlor ^{14}C-labeled in two positions did not show mineralization in subsurface samples (Konopka and Turco, 1991). The apparent incomplete degradation of complex xenobiotics in aquifer systems needs further investigation, from both an environmental fate perspective and from a potential toxicological perspective, assuming that persistent metabolites can be recovered and identified.

D. Anaerobic Conditions

Anaerobic conditions develop in soils and in subsurface materials for varying periods of time in many different agricultural regions. Anaerobic conditions are also commonly found in freshwater aquatic sediments and in riparian environments. Anaerobiosis indicates a lack of free oxygen, but this restriction covers a variety of environments ranging from nitrate- and sulfate-reducing redox conditions (220 to -150 mv redox potential) to methanogenic (< -250 mv) conditions. Anaerobic conditions form in saturated soils where oxygen diffusion becomes restricted. Anaerobic conditions develop seasonally in parts of poorly-drained soil profiles in soils that are generally regarded as aerobic. Even aerobic soils may have anaerobic microsites (Tiedje et al., 1984). Organic substrates serve two basic purposes in soils relative to the extent of anaerobic environments. Organic materials act as substrates for aerobic reactions which deplete oxygen conditions and they serve as substrates for fermentative metabolism under anaerobic conditions. A variety of aromatic substrates are metabolized by nitrate- and sulfate-reducing bacteria. Metabolism of benzoates, phenols, and cresol proceed through similar reductive steps to produce reduced rings (e.g. methylcycloheanone, cyclohexanone), which are then metabolized in several steps to aliphatic acids (Evans and Fuchs, 1988). Alternatively, hydroxylation with water as a source of oxygen, such as the oxidation of toluene to cresol may proceed ring cleavage (Grbic-Galic, 1990).

1. Dehalogenation Pathways

Dehalogenation of aromatic compounds is a common reaction observed under anaerobic conditions, which has been recently reviewed (Kuhn and Suflita, 1989a). This reaction has been observed for a variety of substrates in stream sediments, soil, and in subsurface sediments. Reductive dehalogenation is a microbial process requiring methanogenic conditions. Halogenated benzoates were ultimately degraded to CH_4 and CO_2 (Suflita et al., 1982). Dechlorination, while reducing potential environmental toxicity, does not always lead to complete metabolism of pesticides in anaerobic systems. Anaerobic enrichment cultures from rice paddy soil metabolized propanil initially to 3,4-dichloro-aniline, which was then dechlorinated to form 3-chloroaniline (Pettigrew et al., 1985). Volatile products ($^{14}CO_2$ or $^{14}CH_4$) were not formed. Stepp et al. (1985) observed that diuron was dechlorinated to form 3-chlorophenyl-1,1-dimethylurea which accumulated in nearly stoichiometric quantities. Further incubation produced other metabolites which were not identified. Other phenylureas were also dechlorinated in a similar way. Alachlor added to anaerobic stream sediment was reductively dechlorinated, with sterilization eliminating the reaction (Bollag et al., 1986). The half-life of alachlor was estimated to be 2.5 weeks and the dechlorinated metabolite accumulated over time.

Dechlorination of 3,4-dichloroaniline was also observed in a methanogenic aquifer sediment with accumulation of 3-chloroaniline (Kuhn and Suflita, 1989b). Dechlorination has also been observed with 2,4-D and 2,4,5-T in pond sediment and methanogenic aquifer sediments. The 2,4-D was degraded within 3 months to 2,4-dichlorophenol, 4-chlorophenol and phenol (Gibson and Suflita, 1986). However, no degradation was observed in a sulfate-reducing (nonmetha-nogenic) aquifer in the same study. It was shown by adding sulfate to the methanogenic aquifer or acetate to the sulfate-reducing aquifer material that sulfate inhibited the dehalogenation reactions. Bromacil, a nitrogen-containing heterocyclic herbicide was converted to 3-sec-butyl-methyluracil, but there was no evidence that this compound degraded further (Adrian and Suflita, 1990). These studies indicate that the microorganisms responsible for the reductive dehalogenation reactions appear to be widely distributed in nature. It also appears that the enzyme or isoenzymes responsible for these reactions have a broad range of substrates that they are able to transform, although individual enrichments of dehalogenating microorganisms may show substrate specificity.

2. Reduction and Other Reactions

Pesticides are also reduced under anaerobic conditions. Trifluralin contains two nitro groups which are sequentially reduced under anaerobic conditions (Parr and Smith, 1973). The resulting metabolite then undergoes N-dealkylation to produce metabolite TR-9. Incubation of autoclaved soil under nitrogen atmosphere did not result in trifluralin degradation, which strongly indicates that the reactions are microbially mediated. The nitro group of pedimethalin was reduced to an amino group in a manner similar to that of trifluralin (Smith et al., 1979).

Flumetsulam degraded under sulfate-reducing and methanogenic conditions by the hydration of the pyrimidine ring to form flumetsulam-hydrate (Wolt et al., 1992). This reaction was slow, with a half-life of flumetsulam in laboratory experiments of 180 days. Although biological degradation was not proven, similar biological reactions have been observed in other systems (Evans and

Fuchs, 1988). Dehalogenation or other degradation products of flumetsulam-hydrate were not observed in significant quantities.

3. Flooded Soils

Flooding of soils produces anaerobic conditions which affect the pathways and rates of pesticide metabolism. However, in most agricultural settings flooding is an intermittent condition. The effects of flooding have been compared to continuously aerobic and cyclic flooding conditions, but no generalized statements can be made. For instance, pendimethalin, a dinitroaniline herbicide, was degraded most rapidly under aerobic conditions and slowest under continuously anaerobic conditions, but the redox changes in flooded treatments were not documented (Barrett and Lavy, 1983). Methoxychlor mineralization was greater under anaerobic and denitrifying conditions followed by aerobic incubations than under continuously aerobic conditions (Fogel et al., 1982). During the anaerobic phase of the soil incubations neither $^{14}CO_2$ or $^{14}CH_4$ were formed. A dechlorinated metabolite was formed during the anaerobic treatment, presumably by the microbially-mediated reductive dechlorination process described earlier. Reductive dechlorination requires methanogenic conditions, which would not be present in many soils, even after temporary flooding. However, these conditions would be found in aquatic sediments, subsoils under aerobic soils, and under flooded crops such as rice.

V. Management Factors Affecting Biodegradation in Soil

Farming practices and soil management affect pesticide degradation in several ways. Herbicide formulations and application techniques are principally oriented towards reducing volatilization and drift and to enhance foliar penetration. Little thought has been given to the effect of pesticide applications on soil persistence until the problem of enhanced biodegradation became more apparent. Clearly, repeated use of herbicides and insecticides with similar structures will increase the likelihood of developing adapted microbial populations. Rotation of pesticides appears to have considerable potential for halting microbial adaptation. Practices such as split applications and control-release formulations seek to prolong the time period where pesticide concentrations are sufficient to maintain activity against the target pest. While these practices change the pattern of field persistence, the pathways and factors controlling the degradation rate are not changed. The possibility of split-applications predisposing soils to enhanced biodegradation seems plausible but has not been investigated. Field persistence may play a role in the choice of herbicides. In some instances a certain compound may be chosen over another with similar pesticidal properties on the basis of longer field persistence. In other instances, rates or choices of chemicals may be determined on the desire to avoid undesirable effects such as herbicide carryover.

The diversity of biodegradative pathways and the distributions of these pathways in different species of soil bacteria and fungi indicates that microorganisms will be able to degrade the majority of applied pesticides under most farm management conditions. For instance, no-till management does not appear to drastically change the overall persistence of atrazine or metribuzin over single growing seasons in comparison to tilled soil (Ghadiri et al., 1984; Locke and Harper, 1991a; Sorenson et al., 1991). Tillage did decrease the mineralization of metribuzin and increased the bound fraction despite higher microbial respiration in the no-till soil (Locke and Harper, 1991b). The real issues related to management are the effect of different farming systems on the movement of agricultural chemicals into surface and ground water and the atmosphere. Solving these issues will require accurate knowledge of how pesticides degrade and the factors most influencing biodegradation rates.

VI. Conclusions and Recommendations for Research

The biodegradation process in soil is the principal means of detoxification of pesticides and their ultimate removal from the environment. The organisms responsible for biodegradation are diverse in their taxonomic and ecological nature, which allows the biodegradation process to occur under a wide range of environmental conditions. Practical interests in the biodegradation process arise from a need to provide consistent pest control without carryover of potentially harmful residues and concern over off-site movement of pesticides and metabolites to surface and ground water. More recently, interest in the bioremediation of pesticide spills and wastes in soil and water has increased. Developing solutions to each of these practical concerns is dependent upon some level of understanding of the physiology and ecology of the microorganisms that degrade pesticides.

The initial phase of pesticide degradation has often been described through the use of empirical first-order rate constants. Considerable variability exists in the degradation rate constants for individual compounds in studies where environmental influences have been removed. Experiments which vary the initial pesticide rate over a broad range of concentrations shows that the actual kinetics are not first-order in many cases. These inconsistencies suggest that microbiological variation in the different soils is important. Alternative kinetic models that are more mechanistic from a microbiological viewpoint are available, but have not been verified to have broad predictive capability in soils. One piece of missing information appears to be the size and activity of the microbial biomass involved in the degradation process. Total microbial biomass may be a satisfactory surrogate measure in some instances, but methods for measuring populations or biomass of degradative microorganisms need further development. Gene probe techniques appear to be a potentially useful tool, but they will require considerably more development to be widely applicable to a large number of pesticides.

Microbial adaptation appears to be an important mechanism in the development of enhanced biodegradation and possibly in the metabolism of trace contaminants in subsoils and aquifers. Possible mechanisms of microbial adaptation appear to have been identified, but our understanding of how enzyme induction and specificity interact with population growth does not yet explain why enhanced biodegradation develops (or fails to develop) in soils. The size and population dynamics of pesticide-degrading microorganisms is completely unknown for many classes of compounds. The role of communities in biodegradation appears to be more common than is often recognized, but the difficulty in isolating and manipulating multispecies assemblages for study has probably slowed research in this area. Adaptation may also be important in deeper subsurface sediments where complex xenobiotics including pesticides appear to be persistent. Substantial microbial populations exist in aquifers, but their activity towards pesticides is not well documented. Whether this is due to reduced microbial diversity or environmental limitations is not clear. The fate of pesticide metabolites in both aerobic and anaerobic subsurface materials is largely unknown at this time.

Another issue that appears to be important is the interactions of sorption, bound residues and microorganisms on long-term persistence and availability of pesticide residues. Adsorption to soil may offer some protection from biodegradation, thus enabling some residues to persist for long periods of time in soils. The possibility that adsorbed or bound residues contribute to the phenomenon of pesticide carryover in farmer's fields or to the contamination of ground water needs further investigation. The issue of long-term persistence of pesticides in the environment is important to water and soil quality. The microbiological processes that ultimately control persistence of organics in the environment deserve continued attention.

References

Adams, C.D. and E.M. Thurman. 1991. Formation and transport of deeethyl-atrazine in the soil and vadose zone. *J. Environ. Qual.* 20:540-547.

Adrian, N.R. and J.M. Suflita. 1990. Reductive dehalogenation of a nitrogen heterocyclic herbicide in anoxic aquifer slurries. *Appl. Environ. Microbiol.* 56:292-294.

Alexander, M. and B.K. Lustigman. 1966. Effect of chemical structure on microbial degradation of substituted benzenes. *J. Agric. Food Chem.* 14:410-413.

Alexander, M. and K.M. Scow. 1989. Kinetics of biodegradation in soil. p. 243-269. In: B.L. Sawhney and K. Brown (eds.) *Reactions and Movement of Organic Chemicals in Soils.* Am. Soc. Agron., Madison, WI.

Allen, R. and A. Walker. 1987. The influence of soil properties on the rates of degradation of metamitron, metazachlor, and metribuzin. *Pestic. Sci.* 18:95-111.

Anderson, J.P.E. 1981. Soil moisture and the rates of biodegradation of diallate and triallate. *Soil Biol. Biochem.* 13:155-161.

Anderson, J.P.E. 1984. Herbicide degradation in soil: influence of microbial biomass. *Soil Biol. Biochem.* 16:483-489.

Audus. 1949. Biological detoxification of 2,4-D. *Plant Soil* 3:170-192.

Barrett, M.R. and T.L. Lavy. 1983. Effects of soil and water content on pedimethalin dissipation. *J. Environ. Qual.* 12:504-507.

Baughman, G.L., D.F. Paris, and W.C. Steen. 1980. Quantitative expression of biotransformation rate. p 105-111. In: A.W. Maki, K.L. Dickinson, and J. Cairns, Jr. (eds.), *Biotransformation and fate of chemicals in the aquatic environment*. Am. Soc. Microbiol., Washington, D.C.

Behki, R.M. and S.U. Kahn. 1986. Degradation of atrazine by *Pseudomonas*: N-dealkylation and dehalogenation of atrazine and its metabolites. *J. Agric. Food Chem.* 34:746-749.

Belasco, I.J. and H.L. Pease. 1969. Investigation of diuron- and linuron-treated soils for 3,3',4,4'-tetrachlorobenzene. *J. Agric. Food Chem.* 17:1414-1417.

Bollag, J.-M. and S.-Y. Liu. 1990. Biological transformation processes of pesticides. p. 169-211. In: H.H. Cheng (ed.), *Pesticides in the soil environment: processes, impacts, and modeling*. Soil Sci. Soc. Am., Madison, WI.

Bollag, J.-M., S.-Y. Liu, and R.D. Minnard. 1980. Cross-coupling of phenolic humus constituents and 2,4-dichlorophenol. *Soil Sci. Soc. Am. J.* 44:52-56.

Bollag, J.-M., L.L. McGahen, R.D. Minard, and S.-Y. Liu. 1986. Bioconversion of alachlor in anaerobic stream sediment. *Chemosphere* 15:153-162.

Bruhn, C., H. Lenke, and H.-J. Knackmuss. 1987. Nitrosubstituted aromatic compounds as nitrogen source for bacteria. *Appl. Environ. Microbiol.* 53:208-210.

Burns, R.G. and J.A. Edwards. 1980. Pesticide breakdown by soil enzymes. *Pestic. Sci.* 11:506-512.

Capriel, P., A. Haisch, and S.U. Khan. 1985. Distribution and nature of bound (nonextractable) residues of atrazine in a mineral soil nine years after herbicide application. *J. Agric. Food Chem.* 33:567-569.

Cook, A.M., C.G. Daughton, and M. Alexander. 1978. Phosphorus-containing pesticide breakdown products: quantitative utilization as phosphorus sources by bacteria. *Appl. Environ. Microbiol.* 36:668-672.

Cook, A.M. and R. Hütter. 1981. *s*-Triazines as nitrogen sources for bacteria. *J. Agric. Food Chem.* 29:1135-1143.

Cook, A.M. and R. Hütter. 1982. Ametryne and prometryne as sulfur sources for bacteria. *Appl. Environ. Microbiol.* 43:781-786.

Cook,, A.M. and R. Hütter. 1984. Deethylsimazine: bacterial dechlorination, deamination, and complete degradation. *J. Agric. Food Chem.* 32:581-585.

Cotterill, E.G. and P.G. Owen. 1989. Enhanced degradation in soil of tri-allate and other carbamate pesticides following application of tri-allate. *Weed Res.* 29:65-68.

Cullimore, R.D. 1981. The enumeration of 2,4-D degraders in Saskatchewan soils. *Weed Sci.* 29:440-443.

Daughton, C.G. and D.P.H. Hsieh. 1977. Parathion utilization by bacterial symbionts in a chemostat. *Appl. Environ. Microbiol.* 34:175-184.

Derbyshire, M.K., J.S. Karns, P.C. Kearney, and J.O. Nelson. 1987. Purification and characterization of an N-methylcarbamate pesticide hydrolyzing enzyme. *J. Agric. Food Chem.* 35:871-877.

Dick, W.A., R.O. Ankumah, G. McClung, and N. Abou-Assaf. 1990. Enhanced degradation of S-ethyl N,N-dipropylcarbamothioate in soil and by an isolated soil microorganism. p. 98-112. In: K.D. Racke and J.R. Coats (eds.), *Enhanced biodegradation of pesticides in the environment.* ACS Symp. Series No. 426, Amer. Chem. Soc., Wash., D.C.

Dzantor, E.K. and A.S. Felsot. 1989. Effects of conditioning, cross-conditioning, and microbial growth on development of enhanced biodegradation of insecticides in soil. *J. Environ. Sci. Health* B24:569-597.

Engelhardt, G., P.R. Wallnöfer, and R. Plapp. 1971. Degradation of linuron and some other herbicides and fungicides by a linuron-inducible enzyme obtained from *Bacillus sphaericus. Appl. Microbiol.* 22:284-288.

Engelhardt, G., P.R. Wallnöfer, and R. Plapp. 1973. Purification and properties of an aryl acylamidase of *Bacillus sphaericus,* catalyzing hydrolysis of various phenylamide herbicides and fungicides. *Appl. Microbiol.* 26:709-718.

Engesser, K.H., R.B. Cain, and H.J. Knackmuss. 1988. Bacterial metabolism of side chain fluorinated aromatics: cometabolism of 3-trifluoromethyl(TFM)-benzoate by *Pseudomonas putida (arvilla)* mt-2 and *Rhodococcus rubropertinctus* N657. *Arch. Microbiol.* 149:188-197.

Evans, W.C. and G. Fuchs. 1988. Anaerobic degradation of aromatic compounds. *Ann. Rev. Microbiol.* 42:289-317.

Felsot, A.S. 1989. Enhanced biodegradation of insecticides in soil: implications for agroecosystems. *Ann. Rev. Entomol.* 34:453-476.

Felsot, A.E., J.V. Maddox, and W. Bruce. 1981. Enhanced microbial degradation of carbofuran in soils with histories of carbofuran use. *Bull. Environ. Contam. Toxicol.* 26:781-788.

Feng, P.C.C. 1991. Soil transformations of acetochlor via glutathione conjugation. *Pestic. Biochem. Physiol.* 40:136-142.

Fogel, S., R.L. Lancione, and A.E. Sewall. 1982. Enhanced biodegradation of methoxychlor in soil under sequential environmental conditions. *Appl. Environ. Microbiol.* 44:113-120.

Forrest, M., K.A. Lord, N. Walker, and H.C. Woodville. 1981. The influence of soil treatments on the bacterial degradation of diazinon and other organophosphorus insecticides. *Environ. Pollut. Ser. A.* 24:93-104.

Frehse, H. and J.P.E. Anderson. 1983. Pesticide residues in soil-problems between concept and concern. p. 23-32. In: J. Miyamato and P.C. Kearney (eds.), *IUPAC Pesticide chemistry, human welfare and the environment, Vol. 4, Pesticide residues and formulation chemistry.* Pergamon Press, Oxford.

Führ, F. and W. Mittelstaedt. 1980. Plant experiments on the bioavailability of unextracted [*carbonyl*-¹⁴C]metabenzthiazuron residues from soil. *J. Agric. Food Chem.* 28:122-125.

Gaynor, J.D. 1992. Microbial hydrolysis of diclofop-methyl in soil. *Soil Biol. Biochem.* 24:29-32.

Ghadiri, H., P.J. Shea, G.A. Wicks, and L.C. Haderlie. 1984. Atrazine dissipation in conventional-till and no-till sorghum. *J. Environ. Qual.* 13:549-552.

Ghiorse, W.C. and J.T. Wilson. 1988. Microbial ecology of the subsurface. *Adv. Appl. Microbiol.* 33:107-172.

Gibson, S.A. and J.M. Suflita. 1986. Extrapolation of biodegradation results to groundwater aquifers: reductive dehalogenation of aromatic compounds. *Appl. Environ. Microbiol.* 52:681-688.

Gordon, A.S. and F.J. Millero. 1985. Adsorption mediated decrease in the biodegradation rate of organic compounds. *Microb. Ecol.* 11:289-298.

Grbic-Galic, D. 1990. Anaerobic microbial transformation of nonoxygenated aromatic and alicyclic compounds in soil, subsurface, and freshwater sediments. pp. 116-189. In: J.-M. Bollag and G. Stotzky (eds.), *Soil biochemistry,* Vol. 6. Marcel-Dekker, New York.

Greer, L.E., J.A. Robinson, and D.R. Shelton. 1992. Kinetic comparison of seven strains of 2,4-dichlorophenoxyacetic acid-degrading bacteria. *Appl. Environ. Microbiol.* 58:1027-1030.

Gustafson, D.I. and L.R. Holden. 1990. Nonlinear pesticide dissipation in soil. A new model based on spatial variability. *Environ. Sci. Technol.* 24:1032-1038.

Hamaker, J.W. and C.A.I. Goring. 1976. Turnover of pesticide residues in soil. p. 219-243. In: D.D. Kaufman, G.G. Still, G.D. Paulson, and S.K. Bandal (eds.), *Bound and conjugated pesticide residues.* ACS Symp. Ser. No. 29, Amer. Chem. Soc., Washington. D.C.

Hance, R.J. 1984. Herbicide residues in soil: some aspects of their behavior and agricultural significance. *Australian Weeds* 3:26-34.

Hassett, J.J. and W.L. Banwart. 1989. The adsorption of nonpolar organics by soils and sediments. p. 31-80. In: B.L. Sawhney and K. Brown (eds.), *Reactions and movement of organic chemicals in soils.* Am. Soc. Agron., Madison, WI.

Helling, C.S. and A.E. Krivonak. 1978. Biological characteristics of bound dinitroaniline herbicides in soils. *J. Agric. Food Chem.* 26:1164-1172.

Hendry, K.M. and C.J. Richardson. 1988. Soil biodegradation of carbofuran and furathiocarb following soil pretreatment with these pesticides. *Environ. Toxicol. Chem.* 7:763-774.

Hill, B.D. and G.B. Schaalje. 1985. A two-compartment model for the dissipation of deltamethrin on soil. *J. Agric. Food Chem.* 33:1001-1006.

Holben, W.E., B.M. Schroeter, V.G.M. Calabrese, R.H. Olsen, J.K. Kukor, V.O. Biederbeck, A.E. Smith, and J.M. Tiedje. 1992. Gene probe analysis of soil microbial populations selected by amendment with 2,4-dichlorophenoxyacetic acid. *Appl. Environ. Microbiol.* 58:3941-3948.

Hovarth, R.S. 1972. Microbial co-metabolism and degradation of organic compounds in nature. *Bacteriol. Rev.* 36:146-155.

Janke, D., W. Ihn, and D. Tresselt. 1989. Critical steps in degradation of chloroaromatics by rhodococci. IV. Detailed kinetics of substrate removal and product formation by resting pre-adapted cells. *J. Basic Microbiol.* 29:305-314.

Jones, S.H. and M. Alexander. 1986. Kinetics of mineralization of phenols in lake water. *Appl. Environ. Microbiol.* 51:891-897.

Jutzi, K., A.M. Cook, and R. Hütter. 1982. The degradative pathway of the s-triazine melamine. *Biochem J.* 208:679-684.

Karns, J.S. S. Duttagupta, and A.M. Chakrabarty. 1983. Regulation of 2,4,5-trichlorophenoxyacetic acid and chlorophenol metabolism in *Pseudomonas cepacia* AC1100. *Appl. Environ. Microbiol.* 46:1182-1186.

Kaufman, D.D. 1967. Degradation of carbamate herbicides in soil. *J. Agric. Food Chem.* 15:582-591.

Kaufman, D.D., and J. Blake. 1973. Microbial degradation of several acetamide, acylanilide, carbamate, toluidine and urea pesticides. *Soil Biol. Biochem.* 5:297-308.

Kearney, P.C. 1967. Influence of physicochemical properties on biodegradability of phenyl carbamates. *J. Agric. Food Chem.* 15:568-571.

Kempson-Jones, G.F. and R.J. Hance. 1979. Kinetics of linuron and metribuzin degradation in soil. *Pestic. Sci.* 10:449-454.

Khan, S.U. 1982. Distribution and characteristics of bound residues of prometryn in an organic soil. *J. Agric. Food Chem.* 30:175-179.

Khan, S.U. and R.M. Behki. 1990. Effects of *Pseudomonas* species on the release of bound ^{14}C residues from soil treated with [^{14}C]atrazine. *J. Agric. Food Chem.* 38:2090-2093.

Khan, S.U., and K.C. Ivarson. 1982. Release of soil bound (nonextractable) residues by various physiological groups of microorganisms. *J. Environ. Sci. Health* B17:737-749.

Khan, S.U. and W.J. Saidak. 1981. Residues of atrazine and its metabolites after prolonged usage. *Weed Res.* 21:9-12.

Khan, S.U. P B. Marriage, and J.W. Saidak. 1976. Persistence and movement of diuron and 3,4-dichloroaniline in an orchard soil. *Weed Sci.* 24:583-586.

Konopka, A., and R. Turco. 1991. Biodegradation of organic compounds in vadose and aquifer sediments. *Appl. Environ. Microbiol.* 57:2260-2268.

Kuhn, E. P. and J.M. Suflita. 1989a. Dehalogenation of pesticides by anaerobic microorganisms in soils and groundwater-a review. p. 111-180. In: B.L. Sawhney and K. Brown (eds.), *Reactions and movement of organic chemicals in soils.* Am. Soc. Agron., Madison, WI.

Kuhn, E.P. and J.M. Suflita. 1989b. Sequential reductive dehalogenation of chloroanilines by microorganisms from a methanogenic aquifer. *Environ. Sci. Technol.* 23:848-852.

Lanzilotta, R.P. and D. Pramer. 1970. Herbicide transformation I. Studies with whole cells of *Fusarium solani. Applied Microbiol.* 19:301-306.

Lappin, H.M., M.P. Greaves, and J. Howard Slater. 1985. Degradation of the herbicide mecoprop [2-(2-methyl-4-chlorophenoxy)propanoic acid] by a synergistic microbial community. *Appl. Environ. Microbiol.* 49:429-433.

Larson, R.J. 1980. Role of biodegradation kinetics in predicting environmental fate. p 67-86. In: A.W. Maki, K.L. Dickson, and J. Cairns, Jr. (eds.), *Biotransformation and fate of chemicals in the aquatic environment.* Am. Soc. Microbiol., Washington, D.C.

Larson, R.J. 1984. Kinetic and ecological approaches for predicting biodegradation rates of xenobiotic organic chemicals in natural ecosystems. In: M.J. Klug and C. A. Reddy (eds.), *Current perspectives in microbial ecology.* Am. Soc. Microbiol., Washington, D.C.

Lee, A. 1984. EPTC (*S*-ethyl *N,N*-dipropylthiocarbamate)-degrading microorganisms isolated from a soil previously exposed to EPTC. *Soil Biol. Biochem.* 16:529-531.

Lee, J.K., F. Führ, and W. Mittelstaedt. 1988. Formation and bioavailability of bentazon residues in a German and Korean agricultural soil. *Chemosphere* 17:441-450.

Liu, S.-Y., R.D. Minard, and J.-M. Bollag. 1987. Soil-catalyzed complexation of the pollutant 2,6-diethylaniline with syringic acid. *J. Environ. Qual.* 16:48-53.

Liu, S.-Y., Z. Zheng, R. Zhang, and J.-M. Bollag. 1989. Sorption and metabolism of metolachlor by a bacterial community. *Appl. Environ. Microbiol.* 55:733-740.

Locke, M.A. 1992. Sorption-desorption kinetics of alachlor. *J. Environ. Qual.* 21:558-566.

Locke, M.A. and S.S. Harper. 1991a. Metribuzin degradation in soil: I.-Effects of soybean residue amendment, metribuzin level, and soil depth. *Pestic. Sci.* 31:221-237.

Locke, M.A. and S.S. Harper. 1991b. Metribuzin degradation in soil: II-Effects of tillage. *Pestic. Sci.* 31:239-247.

MacRae, I.C. 1989. Microbial metabolism of pesticides and structurally related compounds. *Rev. Environ. Contam. Toxicol.* 109:1-87.

Marty, J.L. and J. Vogues. 1987. Purification and properties of a phenylcarbamate herbicide degrading enzyme of *Pseudomonas alcaligines* isolated from soil. *Agric. Biol. Chem.* 51:3287-3294.

McCall, P.J. and G.L. Agin. 1985. Desorption kinetics of picloram as affected by residence time in the soil. *Environ. Toxicol. Chem.* 4:37-44.

Moore, J.K., H.D. Braymer, and A.D. Larson. 1983. Isolation of a *Pseudomonas* species which utilizes the phosphate herbicide glyphosate. *Appl. Environ. Microbiol.* 46:316-320.

Moorman, T.B. 1988. Population of EPTC-degrading microorganisms in soils with accelerated rates of EPTC degradation. *Weed Sci.* 36:96-101.

Moorman, T.B. 1990. Adaptation of microorganisms in subsurface environments. p. 167-180. In: *Enhanced biodegradation of pesticides in the environment.* ACS Symp. Ser. No. 426, Amer. Chem. Soc., Washington, D.C.

Moorman, T.B., M.W. Broder, and W.C. Koskinen. 1992. Kinetics of EPTC biodegradation and effects of the inhibitor dietholate in solids. *Soil Biol. Biochem.* 24:121-127.

Moorman, T.B. and S.S. Harper. 1989. Transformation and mineralization of metribuzin in surface and subsurface horizons of a Mississippi Delta soil. *J. Environ. Qual.* 18:302-306.

Mueller, J.G. H.D. Skipper, and E.L. Kline. 1988. Loss of butylate-utilizing ability by a *Flavobacterium. Pestic. Biochem. Physiol.* 32:189-196.

Mueller, J.G., H.D. Skipper, E.G. Lawrence, and E.L. Kline. 1989. Bacterial stimulation by carbamothioate herbicides. *Weed Sci.* 37:424-427.

Mueller, T.C., T.B. Moorman, and C.E. Snipes. 1992. Effect of soil depth on fluometuron sorption and degradation. *J. Agric. Food Chem.* 40:2517-2522.

Mulbry, W.W., and R.W. Eaton. 1991. Purification and characterization of the N-methylcarbamate hydrolase from *Pseudomonas* strain CRL-OK. *Appl. Environ. Microbiol.* 57:3679-3682.

Munnecke, D.M., L.M. Johnson, H.W. Talbot, and S. Barik. 1982. Microbial metabolism and enzymology of selected pesticides. pp. 1-32. In: A. M. Chakrabarty (ed.), *Biodegradation and detoxification of environmental pollutants.* CRC Press, Boca Raton, FL.

Nakamura, T., K. Mochida, W.X. Li, and Y. Ozoe. 1992. Isolation of aryl acylamidase-producing bacteria and some properties of the extracellular enzymes. *J. Pestic Sci.* 17:99-106.

Nash, R.G. 1988. Dissipation from soil. p. 131-169. In: R. Grover (ed.), *Environmental chemistry of herbicides,* Vol. I. CRC Press, Boca Raton, FL.

Nelson, J.E., W.F. Meggitt, and D. Penner. 1983. Fractionation of residues of pendimethalin, trifluralin and oryzalin during degradation in soil. *Weed Sci.* 31:68-75.

Nelson, L.M., B. Yaron, and P.H. Nye. 1982. Biologically-induced hydrolysis of parathion in soil: kinetics and modelling. *Soil Biol. Biochem.* 14:223-227.

Novick, N.J., R. Mukherjee, and M. Alexander. 1986. Metabolism of alachlor and propachlor in suspensions of pretreated soils and in samples from groundwater aquifers. *J. Agric. Food Chem.* 34:721-725.

Obrigawitch, T., A.R. Martin, and F.W. Roeth. 1983. Degradation of thiocarbamate herbicides in soils exhibiting rapid EPTC breakdown. *Weed Sci.* 31:187-192.

Obrigawitch, T., R. G. Wilson, A. R. Martin, and F. W. Roeth. 1982. The influence of temperature, moisture, and prior EPTC application on the degradation of EPTC in soils. *Weed Sci.* 30:175-181.

Ogram, A.V., R.E. Jessup, L.-T. Ou, and P.S.C. Rao. 1985. Effects of sorption on biological degradation rates of (2,4-dichlorophenoxy)acetic acid in soils. *Appl. Environ. Microbiol.* 49:582-587.

Ou, L.-T. 1984. 2,4-D degradation and 2,4-D degrading microorganisms in soils. *Soil Sci.* 137:100-107.

Paris, D.F., W.C. Steen, G.L. Baughman, and J.T. Barnett, Jr. 1981. Second-order model to predict microbial degradation of organic compounds in natural waters. *Appl. Environ. Microbiol.* 41:603-609.

Parr, J.F. and S. Smith. 1973. Degradation of trifluralin under laboratory conditions and soil anaerobiosis. *Soil Sci.* 115:55-63.

Pettigrew, C.A., M.J.B. Paynter, and N.D. Camper. 1985. Anaerobic microbial degradation of the herbicide propanil. *Soil Biol. Biochem.* 17:815-818.

Pignatello, J.J. 1989. Sorption dynamics of organic compounds in soils. p. 45-80. In: B.L. Sawhney and K. Brown (eds.), *Reactions and movement of organic chemicals in soils*. Am. Soc. Agron., Madison, WI.

Pignatello, J.J. 1990a. Slowly reversible sorption of aliphatic hydrocarbons in soils. I. Formation of residual fractions. *Environ. Toxicol. Chem.* 9:1107-1115.

Pignatello, J.J. 1990b. Slowly reversible sorption of aliphatic hydrocarbons in soils. II. Mechanistic aspects. *Environ. Toxicol. Chem.* 1117-1126.

Pimental, D. and L. Levitan. 1986. Pesticides: amounts applied and amounts reaching pests. *Bioscience* 36:86-91.

Pothuluri, J.V., T.B. Moorman, D.C. Obenhuber, and R.D. Wauchope. 1990. Aerobic and anaerobic degradation of alachlor in samples from a surface to groundwater profile. *J. Envrion. Qual.* 19:525-530.

Racke, K.D. and J.R. Coats. 1987. Enhanced degradation of isofenphos by soil microorganisms. *J. Agric. Food Chem.* 35:94-99.

Racke, K.D. and E.P. Lichtenstein. 1985. Effects of soil microorganisms on the release of bound ^{14}C residues from soils previously treated with [^{14}C] parathion. *J. Agric. Food Chem.* 33:938-943.

Read, D.C. 1983. Enhanced microbial degradation of carbofuran and fensulfo-thion after repeated applications to acid mineral soil. *Agric. Ecosystems Environ.* 10:37-46.

Reichel, H., H.D. Sisler, and D.D. Kaufman. 1991. Inducers, substrates, and inhibitors of a propanil-degrading amidase of *Fusarium oxysporum*. *Pestic. Biochem. Physiol.* 39:240-250.

Roeth, F.W. 1986. Enhanced herbicide degradation in soil with repeat application. *Rev. Weed Sci.* 2:45-65.

Rosenberg, A. and M. Alexander. 1980. Microbial metabolism of 2,4,5-trichlorophenoxyacetic acid in soil, soil suspensions, and axenic culture. *J. Agric. Food Chem.* 28:297-302.

Ruggiero, P. J.M. Sarkar, and J.-M. Bollag. 1989. Detoxification of 2,4-dichlorophenol by a laccase immobilized on soil or clay. *Soil Sci.* 147:361-370.

Sariaslani, F.S. 1991. Microbial cytochromes *P*-450 and xenobiotic metabolism. *Adv. Appl. Microbiol.* 36:133-178.

Saxena, A. and R. Bartha. 1983. Microbial mineralization of humic acid-3,4-dichloroaniline complexes. *Soil Biol. Biochem.* 15:59-62.

Schiavon, M. 1988. Studies of the leaching of atrazine, of its chlorinated derivatives, and of hydroxyatrazine from soil using ^{14}C ring-labeled compounds under outdoor conditions. *Ecotoxicol. Environ. Safety* 15:46-54.

Schmidt, S.K. and M.J. Gier. 1989. Dynamics of microbial populations in soil: indigenous microorganisms degrading 2,4-dinitrophenol. *Microb. Ecol.* 18:285-296.

Schmidt, S.K., and M.J. Gier. 1990. Coexisting bacterial populations responsible for multiphasic mineralization kinetics in soil. *Appl. Environ. Microbiol.* 56:2692-2697.

Scow, K.M. and M. Alexander. 1992. Effect of diffusion on the kinetics of biodegradation: experimental results with synthetic aggregates. *Soil Sci. Soc. Am. J.* 56:128-134.

Scow, K.M., and J. Hutson. 1992. Effect of diffusion and sorption on the kinetics of biodegradation: theoretical considerations. *Soil Sci. Soc. Am. J.* 56:119-127.

Scow, K.M., S. Simkins, and M. Alexander. 1986. Kinetics of mineralization of organic compounds at low concentrations in soil. *Appl. Environ. Microbiol.* 51:1028-1035.

Sharp, D.B. 1988. Alachlor. p. 301-333. In: P.C. Kearney and D.D. Kaufman (eds.), *Herbicides chemistry, degradation and mode of action*, Vol. 3. Marcel Dekker, New York.

Shelton, D.R. and T.B. Parkin. 1991. Effect of soil moisture on sorption and biodegradation of carbofuran in soil. *J. Agric. Food Chem.* 39:2063-2068.

Simkins, S. and M. Alexander. 1984. Models for mineralization kinetics with the variables of substrate concentration and population density. *Appl. Environ. Microbiol.* 47:1299-1306.

Simkins, S. and M. Alexander. 1985. Nonlinear estimation of the parameters of Monod kinetics that best describe mineralization of several substrate concentrations by dissimilar bacterial densities. *Appl. Environ. Microbiol.* 50:816-824.

Simon, L., M. Spiteller, A. Haisch, and P.R. Wallnöfer. 1992. Influence of soil properties on the degradation of the nematacide fenamiphos. *Soil Biol. Biochem.* 24:769-773.

Skipper, H.D., E.C. Murdock, D.T. Gooden, J.P. Zublena, and M.A. Amakiri. 1986. Enhanced herbicide biodegradation in South Carolina soils previously treated with butylate. *Weed Sci.* 34:558-563.

Slater, J.H. and D. Lovatt. 1984. Biodegradation and significance of microbial communities. p. 439-485. In: D.T. Gibson (ed.), *Microbial degradation of organic compounds.* Marcel Dekker, New York.

Smith, A.E. 1978. Relative persistence of di- and tri-chlorophenoxyacetic acid herbicides in Saskatchewan soils. *Weed Res.* 18:275-279.

Smith, A.E. and A.J. Aubin. 1992. Degradation of the sulfonylurea herbicide [^{14}C]amidosulfuron (HOE 075032) in Saskatchewan soils under laboratory conditions. *J. Agric. Food Chem.* 40:2500-2504.

Smith, A.E. and D.C.G. Muir. 1984. Determination of extractable and nonextractable radioactivity from small field plots 45 and 95 weeks after treatment with [^{14}C]dicamba, (2,4-dichloro[^{14}C]phenoxy)acetic acid, [^{14}C]triallate, and [^{14}C]trifluralin. *J. Agric Food Chem.* 32:588-593.

Smith, R.H., J.E. Oliver, and W.R. Lusby. 1979. Degradation of pendimethalin and its *N*-nitroso and *N*-nitro derivatives in anaerobic soil. *Chemosphere* 12:855-861.

Somasundaram, L., J.R. Coats, and K.D. Racke. 1989. Degradation of pesticides in soil as influenced by the presence of hydrolysis metabolites. *J. Environ. Sci. Health* B24:457-478.

Sorenson, B.A., P.J. Shea, and F.W. Roeth. 1991. Effects of tillage, application time and rate on metribuzin degradation. *Weed Res.* 31:333-345.

Steinberg, S.M., J.J. Pignatello, and B. Sawhney. 1987. Persistence of 1,2-dibromomethane in soils: entrapment in intraparticle micropores. *Environ. Sci. Technol.* 21:1201-1208.

Stepp, T.D., N.D. Camper, and M.J.B. Paynter. 1985. Anaerobic microbial degradation of selected 3,4-dihalogenated aromatic compounds. *Pestic. Biochem. Physiol.* 23:256-260.

Suflita, J.M., A. Horowitz, D.R. Shelton, and J.M. Tiedje. 1982. Dehalogenation: a novel pathway for the anaerobic biodegradation of haloaromatic compounds. *Science* 218:1115-1116.

Tal, A., B. Rubin, J. Katan, and N. Aharonson. 1989. Fate of ^{14}C-EPTC in a soil exhibiting accelerated degradation of carbamothioate herbicides and its control. *Weed Sci.* 37:434-439.

Tal, A., B. Rubin, J. Katan, and N. Aharonson. 1990. Involvement of microorganisms in accelerated degradation of EPTC in soil. *J. Agric. Food Chem.* 38:1100-1105.

Tam, A.C., R.M. Behki, and S.U. Khan. 1987. Isolation and characterization of an *s*-ethyl-*N,N*-dipropylthiocarbamate-degrading *Arthrobacter* strain and evidence for plasmid-associated *s*-ethyl-*N,N*-dipropylthiocarbamate degradation. *Appl. Environ. Microbiol.* 53:1088-1093.

Tiedje, J.M., A.J. Sexstone, T.B. Parkin, and N.P. Revsbech. 1984. Anaerobic processes in soil. *Plant Soil* 76:197-212.

Turco, R.F. and A. Konopka. 1990. Biodegradation of carbofuran in enhanced and nonenhanced soils. *Soil Biol. Biochem.* 22:195-201.

Villarreal, D.T., R.F. Turco, and A. Konopka. 1991. Propachlor degradation by a soil bacterial community. *Appl. Environ. Microbiol.* 57:2135-2140.

Walker, A. 1987. Further observations on the enhanced degradation of iprodione and vinclozinone in soil. *Pestic. Sci.* 21:219-231.

Walker, A. and A. Barnes. 1981. Simulation of herbicide persistence in soil: a revised computer model. *Pestic. Sci.* 12:123-132.

Walker, A., P.A. Brown, and P.R. Mathews. 1985. Persistence and phytotoxicity of napropamide residues in soil. *Ann. Appl. Biol.* 106:323-333.

Winkelmann, D.A. and S.J. Klaine. 1991. Degradation and bound residue formation of atrazine in a western Tennessee soil. *Environ. Toxicol. Chem.* 10:335-345.

Wolf, D.C., and J.P. Martin. 1975. Microbial decomposition of ring-^{14}C atrazine, cyuranic acid, and 2-chloro-4,6-diamino-s-triazine. *J. Environ. Qual.* 4:134-139.

Wolt, J.D., J.D. Schwake, F.R. Batzer, S.M. Brown, L.H. McKendry, J.R. Miller, G.A. Roth, M.A. Stanga, D. Portwood, and D.L. Holbrook. 1992. Anaerobic aquatic degradation of flumetsulam [*N*-(2,6-difluorophenyl)-5-methyl[1,2,4]triazolo[1,5-a]pyrimidine-2-sulfon amide]. *J. Agric. Food Chem.* 40:2302-2308.

Wszolek, P.C. and M. Alexander. 1979. Effect of desorption rate on the biodegradation of n-alkylamines bound to clay. *J. Agric. Food Chem.* 27:410-414.

Yaron, B., Z. Gertsl, and W. F. Spencer. 1985. Behavior of herbicides in irrigated soils. *Adv. Soil Sci.* 3:121-211.

Zeyer, J. and P.C. Kearney. 1982a. Microbial degradation of para-chloroaniline as sole carbon and nitrogen source. *Pestic. Biochem. Physiol.* 17:215-223.

Zeyer, J. and P.C. Kearney. 1982b. Microbial metabolism of propanil and 3,4-dichloroaniline. *Pestic. Biochem. Physiol.* 17:224-231.

Zeyer, J., H.P. Kocher, and K.N. Timmis. 1986. Inluence of para-substituents on the oxidative metabolism of *o*-nitrophenols by *Pseudomonas putida* B2. *Appl. Environ. Microbiol.* 52:334-339.

Appendix A: Chemical Names of Pesticides and Metabolites

Acetochlor: 2-chloro-*N*-(ethoxymethyl)-*N*-(2-ethyl-6-methylphenyl)acetamide

Alachlor: 2-chloro-*N*-(2,6-diethylphenyl)-*N*-(methoxymethyl)acetamide

Aldicarb: 2-methyl-2-(methylthio)propionaldehyde *O*-[(methylamino) carbonyl]oxime

Ametryn:*N*-ethyl-*N'*-(1-methylethyl)-6-(methylthio)-1,3,5-triazine-2,4-diamine

Amidosulfuron: 3-(4,6-dimethoxypyrimidin-2-yl)-1-(*N*-methyl-*N*-methylsulfonyl) aminosulfonylurea

Atrazine: 6-chloro-*N*-ethyl-*N'*-(1-methylethyl)-1,3,5-triazine-2,4-diamine
  Deethylatrazine: 2-chloro-4-amino-6-isopropylamino-*s*-triazine
  Deisopropylatrazine: 2-chloro-4-ethylamino-6-amino-*s*-triazine
Bromacil: 5-bromo-6-methyl-3-(1-methylpropyl)-2,4(1*H*,3*H*) pyrimidinedione
Butylate: *S*-ethyl bis(2-methylpropyl)carbamothioate
Carbaryl: 1-napthalenol methylcarbamate
Carbofuran: 2,3-dihydro-2,2-dimethyl-7-benzofuranyl methylcarbamate
CIPC (chlorpropham): 1-methylethyl 3-chlorophenylcarbamate
Diazinon: *O,O* diethyl *O*-(6-methyl-2-(1-methylethyl) 4-pyrimidinyl
  phosphorothioate
Diclofop-methyl:(±2-[4-(2,4-dichlorophenoxy)phenoxy]propanoic acid)methyl
  ester
2,4-D: (2,4-dichlorophenoxy)acetic acid
Diallate: *S*-(2,3-dichloro-2-propenyl) bis(1-methylethyl)carbamothioate
EPTC: *S*-ethyl dipropylcarbamothioate
Fenamiphos: ethyl 3-methyl-4-(methylthio)phenyl (1-
  methylethyl)phosphoramidate
Fensulfothion: *O,O*-diethyl *O*-[4-(methylsulfinyl)phenyl]phosphorothioate
Flumetsulam: *N*-(2,6-difluorophenyl)-5-methyl[1,2,4]triazolo[1,5-*a*]pyrimidine-2-
  sulfonamide
Fluometuron: *N,N*-dimethyl-*N'*-[3-(trifluoromethyl)-phenyl]urea
  Desmethylfluometuron: *N*-methyl-*N'*-[3-(trifluoromethyl)-phenyl]urea
Glyphosate: *N*-(phosphonomethyl)glycine
IPC: isopropyl *N*-phenylcarbamate
Iprodione:3-(3,5-dichlorophenyl)-*N*-(1-methylethyl)-2,4dioxo-1-
  imidazolidinecarboxamide
Isofenphos: 1-methylethyl 2-[[ethoxy[(1-methylethyl)amino] phosphinothioyl]
  oxy]benzoate
Linuron: *N'*-(3,4-dichlorophenyl)-*N*-methoxy-*N*-methylurea
Malathion: diethyl mercaptosuccinate *O,O*-dimethyl phosphorothioate
Mecoprop: (±)-2-(4-chloro-2-methylphenoxy)propanoic acid
Metamitron: 4-amino-3-methyl-6-phenyl-1,2,4-triazin-5(4H)-one
Metazachlor: 2-chloro-*N*-(2,6-dimethylphenyl)-*N*-(1*H*-pyrazol-1-ylmethyl)-
  acetamide
Methabenzthiazuron: *N*-(2-benzothiazolyl)-*N,N'*-dimethylurea
Methoxychlor: 1,1'-(2.2.2-trichloroethylidene)-bis[4-methoxybenzene]
Metolachlor: 2-chloro-*N*-(2-ethyl-6-methylphenyl)-*N*-(2-methoxy-1-methylethyl)-
  acetamide
Metribuzin: 4-amino-6-(1,1-dimethylethyl)-3-(methylthio)-1,2,4-triazin-5(4*H*)-one
Napropamide: *N,N*-diethyl-2-(1-napthylenyloxy)propanamide
Parathion: *O,O*-diethyl *O*-(4-nitrophenyl)phosphorothioate
Pendimethalin: *N*-(1-ethylpropyl)-3,4-dimethyl-2,6-dinitrobenzeneamine
Picloram: 4-amino-3,5,6-trichloro-2-pyridinecarboxylic acid
Prometryn: *N,N'*-bis(1-methylethyl)-6-(methylthio)-1,3,5-triazine-2,4-diamine

Propachlor: 2-chloro-*N*-(1-methylethyl)-*N*-phenylacetamide
Propanil: *N*-(3,4-dichlorophenyl)propanamide
Propham: 1-methylethyl phenylcarbamate
Simazine: 6-chloro-*N*,*N'*-diethyl-1,3,5-triazine-2,4-diamine
   Deethylsimazine: 6-chloro-*N*-ethyl-1,3,5-triazine-2,4-diamine
2,4,5-T: (2,4,5-trichlorophenoxy)acetic acid
Triallate: *S*-(2,3,3-trichloro-2-propenyl) bis(1-methylethyl)carbamothioate
Trifluralin: 2,6-dinitro-*N*,*N*-dipropyl-4-(trifluoromethyl)benzenamine
   TR-9: 5-(trifluoromethyl)-1,2,3-benzenetriamine
Vernolate: *S*-propyl dipropylcarbamothioate
Vinclozolin: 3-(3,5-dichlorophenyl)-5-ethenyl-5-methyl-2,4-oxazolidinedione

Index